Henry Leffmann, William Beam

Examination of Water for Sanitary and Technical Purposes

Henry Leffmann, William Beam

Examination of Water for Sanitary and Technical Purposes

ISBN/EAN: 9783744649551

Printed in Europe, USA, Canada, Australia, Japan

Cover: Foto ©berggeist007 / pixelio.de

More available books at **www.hansebooks.com**

EXAMINATION OF WATER

FOR

SANITARY AND TECHNICAL PURPOSES.

BY

HENRY LEFFMANN, M. D., Ph.D.,

PROFESSOR OF CHEMISTRY IN THE WOMAN'S MEDICAL COLLEGE OF PENNSYLVANIA,
IN THE PENNSYLVANIA COLLEGE OF DENTAL SURGERY AND IN THE
WAGNER FREE INSTITUTE OF SCIENCE; PATHOLOGICAL
CHEMIST TO THE JEFFERSON MEDICAL
COLLEGE HOSPITAL,

AND

WILLIAM BEAM, M.A.,

DEMONSTRATOR OF CHEMISTRY IN THE PENNSYLVANIA COLLEGE OF DENTAL
SURGERY; ASSOCIATE OF THE SOCIETY OF PUBLIC ANALYSTS
OF GREAT BRITAIN; FORMERLY CHIEF
CHEMIST, B. & O. R. R.

PHILADELPHIA:
P. BLAKISTON, SON & CO.,
1012 WALNUT STREET.
1889.

PREFACE.

The scope of the present work is sufficiently indicated in the title. In its compilation we have endeavored to select processes which are trustworthy and practicable, and to describe them concisely and clearly. Independently of the treatment which the topic has received in works on general hygiene and sanitary chemistry, there are several special manuals at hand, but they are limited in scope, and contain much matter which does not bear on the problems that confront the sanitarian and industrial chemist, devoting much space to the description of processes that are not generally employed, and to the presentation of the details of a local professional quarrel.

Those already familiar with the literature of this subject, will notice the omission of certain processes which have long held a prominent place. Among these may be mentioned the use of soap solution for the determination of hardness. Hehner has shown that this is frequently inaccurate, and has devised the method which we describe. Allen also takes this view, and has proposed a similar method. In view of the observations of Percy Smith and others as

In the description of the general quantitative analysis, we have followed to a large extent the methods indicated by Fresenius, selecting those best adapted to technical purposes, and in many of the sanitary examinations, we have given the processes approved by the Society of Public Analysts, of Great Britain.

It has not been deemed necessary to attempt a description of the examination of water sediments by the microscope. Such examinations are occasionally interesting from a biological point of view, but results of positive sanitary value are rarely, if ever, attained. A brief notice has been given of the bacteriological examinations, but the employment of this is a matter of special skill, and it can scarcely be safely applied except by those who have had some previous training in the manipulations involved.

Care has been taken to have the nomenclature and notation conform to the most advanced and precise systems in vogue; and to assist in precision in this respect we have furnished a set of labels adapted to the reagents used.

HENRY LEFFMANN,
WILLIAM BEAM.

715 Walnut Street.
April, 1889.

CONTENTS.

HISTORY OF NATURAL WATERS.
    Classification — Rain Water—Surface Water — Subsoil Water—Deep Water, ... 9–15

ANALYTICAL OPERATIONS.
    Sanitary Examinations :—
        Collection and Preliminary Examination—Total Solids—Chlorine—Nitrogen in Ammonium Compounds and Organic Matter—Nitrogen as Nitrates—Nitrogen as Nitrites — Oxygen-consuming Power — Phosphates—Dissolved Oxygen—Poisonous Metals, ... 16–49
    Technical Examinations :—
        General Quantitative Analysis—Spectroscopic Analysis—Specific Gravity, ... 50–65

INTERPRETATION OF RESULTS.
    Statement of Analysis :— 66, 67
    Sanitary Applications :— 68–77
        Action of Water on Lead—Living Organisms in Water—Purification of Drinking Water—Identification of the Source of Water, ... 79–88
    Technical Applications :—
        Boiler Waters—Purification of Boiler Waters, . . 89–96

ANALYTICAL DATA.
    Conversion Table—Oxygen Dissolved in Water—Rain and Subsoil Waters — Schuylkill Water — Artesian Waters—City Supplies, ... 97–102

"If thou couldst, doctor, cast
The water of my land . . . ."

Examination of Water.

History of Natural Waters.

Pure water is an artificial product of the laboratory. Natural waters always contain foreign matters in solution and suspension, varying from mere traces to very large proportions. The properties, effects and uses of water are considerably modified by these ingredients, and the object of analysis is to ascertain their character and amount; and since these are largely dependent on the history of the water, a classification based on this will be convenient. We may distinguish four classes of natural waters:—

Rain Water.—Water precipitated from the atmosphere under any conditions, and therefore including dew, frost, snow and hail.

Surface Water.—All collections of water in streams, seas, lakes, or ponds.

Subsoil Water.—Water percolating or flowing through soil or rock at moderate distance below the surface, and derived in large part from the rain or surface water of the district.

Deep Water.—Water accumulated at considerable depth below the surface, from which the soil water of the district has been excluded by difficultly permeable strata (artesian water).

Rain Water, when gathered in the open country and in the latter period of a long rain or snow, is the purest form of natural water. When collected directly, it contains but little solid matter, this consisting principally of ammonium compounds and particles of organic matter, living and dead, gathered from the atmosphere. In districts near the sea an appreciable amount of chlorides will be present. It is obvious that a prolonged rain will wash out the air, but since storms are usually attended by wind, fresh portions of air are continually flowing in, and thus the water never becomes perfectly pure.

Surface Water.—Rain water in part flows off on the surface of the ground, and at once gains decidedly in the proportion of suspended and dissolved matters. The former are found in large amount when the rainfall is profuse. The wearing action of water is dependent on the amount and character of these suspended materials. From the higher levels of a watershed, the streams, more or less in the form of torrents, gather into larger currents, and reaching lower levels become slower in movement, and deposit much of the suspended matter. By admixture of the waters from widely separated districts the character and amount of the dissolved matters are much modified. An action of this kind is seen in the watershed of the Schuylkill River. This stream rises in the anthracite coal region of Pennsylvania, and receiving much refuse mine water becomes impregnated with iron salts and free mineral acid, being then quite unsuitable for drinking or manufacturing purposes. In its course of about one hundred miles, it passes over an extensive limestone district, and receives several large streams highly charged with calcium carbonate. The result is a neutralization of the acid, and a precipitation of the iron and much of the calcium. The river becomes purer, and at its junction with the Delaware River at Philadelphia, it contains neither free sulphuric nor hydrochloric acid, only traces of iron, and but a small amount of calcium sulphate. In this manner there is produced a soft water, superior to that of the river near its source, or to the hard waters of the middle Schuylkill region.

It is obviously impossible to establish close standards of composition for surface waters. In the case of waters wholly derived from rain, falling on the surface of undisturbed, unpopulated territory, the amount of dissolved solids will be small, and will consist principally of carbonates and sulphates. The water of lakes and rivers is, however, always in part derived from springs, which may proceed from great depths, and thus introduce many substances not soluble in surface water, nor derivable from the soil of the district.

The exposure to light and air which surface water undergoes results in the absorption of oxygen and loss of carbonic acid, together with the oxidation of the organic matter. The diminution of the rapidity of the current permits the deposition of the suspended matters, and this occurs especially as the river approaches the sea, not only from the retarding influence of the tidal wave, but from the precipitating action of the salt water. The investigations of Carl Barus, published in Bulletin No. 36, U. S. Geological Survey, have shown the decided influence of sodium chloride in accelerating the subsidence of fine particles.

Subsoil Water. Water which penetrates into the soil passes to various depths, according to the porosity and arrangement of the strata. As a rule, it descends until it reaches but slightly pervious formations, upon the level of which it accumulates. In the upper layer of soil it dissolves mineral and organic ingredients, and becomes impregnated with micro-organisms, through the agency of which the organic matter undergoes important transformations. The water constantly accumulating gradually flows along the incline of the impervious stratum on which it rests, or through its fissures, and may either pass downward or emerge in the form of a spring.

The proportion of water which may be held by any rock or soil is often much larger than would be at first supposed. T. Sterry Hunt states that a square mile of sandstone 100 feet thick will contain water sufficient to sustain a flow of a cubic foot a minute for more than thirteen years.

Much difference is observed in the composition of subsoil waters, but as a general rule they contain but a limited amount of mineral substances, and a very small proportion of organic matter. In populated districts, however, a marked change is produced through the influence of the animal and vegetable products with which the soil is impregnated. These are in various stages of decomposition, and are either freely soluble or easily suspended in the water. It is especially the organic matter containing nitrogen that is of importance. In this class belong all those tissues that are intimately associated with vital action; also many characteristic excretory products. These bodies are mostly unstable, and as soon as their vitality ceases begin to decompose, partly by oxidation, partly by splitting up into simpler forms; these changes being in many cases brought about by micro-organisms. Among the products noticed in the early stages of such decay, are substances which possess close analogies to the organic bases or alkaloids, but more susceptible of decomposition. They are generally present in minute amount, but are not infrequently very active in their physiological effect. From the most recent researches it seems probable that the pathogenic power of many microorganisms rests not upon any mechanical or other action of the germs themselves, but upon the alkaloidal principles which they produce and excrete. As a group these bodies are known as the "ptomaines." Nitrogen is an invariable ingredient. The ultimate results of these processes of decomposition depend largely on circumstances. When organic matters containing nitrogen are subjected to the action of powerful oxidizing agents, such as alkaline potassium permanganate or chromic acid, some of the nitrogen is converted into ammonia. The same result occurs in all waters, but a considerable portion of the organic matter may also suffer further oxidation and in association with the mineral substances present form nitrates and nitrites, especially the former. This conversion is called "nitrification." The conditions under which it occurs have been carefully studied by Warrington, Munro and others.

Nitrification undoubtedly takes place under the influence of certain microbes the habitat of which does not extend for more than a few yards below the surface of the soil. Rain water caught from the clouds does not contain them. The nitrifying action may be exerted either upon the organic matter or upon the ammonia which is formed from it. The presence of some substance capable of neutralizing acids is necessary. In soils, calcium carbonate or magnesium carbonate usually fulfills this function. Nitrates are the typical results of this action, nitrites being present at any given time only in small quantities. Denitrification, that is, the reduction of nitrites and nitrates to ammonia, takes place under the influence of microbes, and is especially apt to occur when considerable quantities of decomposing organic matter are introduced. A partial reduction sometimes occurs, and a notable proportion of nitrites is found, but in the presence of actively decomposing organic matter, such as that in sewage, a complete reduction, even to the liberation of nitrogen, may occur.

Deep Water.—Water which penetrates the fissures of the fundamental rock formations may pass to great depths, and by following the lines of the lowest and least permeable strata may be transported to points far removed from those at which it was originally collected. The chemical changes which it induces include most of those which take place at higher points, but the increase of pressure and temperature confers much increased power. Carbonic acid will accumulate under conditions favorable to the solution of calcium, magnesium and iron carbonates, and iron and manganese oxides may be converted into carbonates and then dissolved. Sulphates are reduced to sulphides, and these subsequently, by the action of carbonic acid, yield hydrogen sulphide. In these actions the organic matter plays an important part, determining the reduction of ferric compounds to ferrous, and also that of the sulphates, and is converted ultimately into ammonium compounds, notable quantities of which are often found in deep waters. Further, it is found that nitrates and nitrites are present only in small amount, except from certain strata rich in organic matter. In some cases the water acquires very high temperature, and dissociation of rocks occurs, with solution of considerable amounts of silicic acid, which is ordinarily but sparingly soluble in water.

Masses of water thus accumulated under heat and pressure may find their way to the surface either through natural fissures, or be reached by borings. The mineral springs, highly charged with solid matters, and the artesian waters, are obtained in this way.

While no absolute unchangeable line can be drawn between deep and subsoil waters, yet it will in most cases be found that the deep water of a given district, whether obtained through natural or artificial channels, will be decidedly different in composition from the subsoil or surface water of the same, and that the rocks passed through in such cases will be characterized by one or more strata, difficultly permeable to water, and therefore preventing direct communication. The characteristic differences between surface, subsoil, and deep waters are clearly indicated in the table of analyses made by us of samples from the Philadelphia district.

The fact that mere depth is not the essential difference between two classes of waters is shown by comparison between the composition of the water from the well at Barren Hill, on the northern border of Philadelphia county, and the deep well at Locust Point, Baltimore. The former is a dug well, 130 feet deep; the latter is an artesian boring of 128 feet, which in its descent passes through four feet of solid rock. The deeper well is evidently supplied by subsoil water. The artesian well, though located 100 yards from a brackish sewage-laden estuary, evidently derives no water from it.

ANALYTICAL OPERATIONS.

SANITARY EXAMINATIONS.

COLLECTION AND PRELIMINARY EXAMINATION OF SAMPLES.

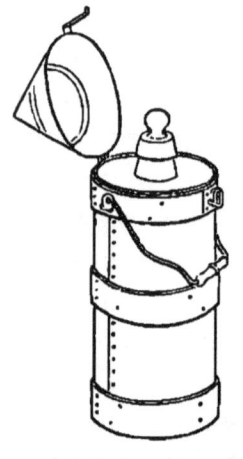

Great care must be taken in collecting water samples, in order to secure a fair representation of the supply and to avoid introduction of foreign matters. The five pint green glass stoppered bottles used for holding acids are suitable for containing the samples. The contents of two such bottles will suffice for most sanitary or technical examinations, and frequently only one will be required. The annexed cut shows a permanently encased bottle, known under the name of the "Penn Demijohn," which we have found very convenient for transportation. It is provided with a hinged lid which can be securely fastened by a padlock. The green glass stoppered bottles may also be fitted to such an arrangement. Stone jugs, casks or metal vessels must not be employed. The bottles used must be thoroughly rinsed several times with the water to be examined, filled quite full, the stopper inserted and tied down, or fastened by stretching over the

PRELIMINARY EXAMINATION OF SAMPLES.

stopper and lip a rubber finger-cot. If corks are used, they should be new and thoroughly rinsed before using. No wax, putty, plaster or similar material should be used.

In taking samples from lakes, slow streams or reservoirs, it is necessary to submerge the bottle so as to avoid collecting any water that has been in immediate contact with the air.

In the examination of public water supplies, the sample should be drawn from a hydrant in direct connection with the main, and not from a cistern, storage tank or dead end of a pipe. In the case of pump-wells, a few gallons of the water should be pumped out before taking the sample, in order to remove that which has been standing in the pipe.

In all cases care should be taken to fill the vessel with as little agitation with air as possible.

It is important that with each sample a record should be made of those surroundings and conditions which might influence the character of the water, particularly in reference to sources of pollution, such as proximity to cesspools, sewers or manufacturing establishments. The character and condition of the different strata of the locality should be noted if possible.

Determinations of nitrogen existing as ammonium compounds, and as organic matter, and of oxygen-consuming power, should be made upon the sample in the original condition, whether turbid or clear, but all other estimations should be made upon the clear liquid. Turbid waters may be clarified by standing or by filtration; for the latter purpose Schleicher & Schüll's extra heavy No. 598 paper is the best. In many cases the sediment cannot be entirely removed by filtration, and subsidence must be resorted to. The use of a small quantity of alum, as described in the section on the purification of water, will often be applicable as a rapid means of clarifying water samples. For the quantitative determination, the sediment from a known volume of the water is collected on a tared filter, dried at 212° F., and weighed.

The water from newly-dug wells is generally turbid and the determinations are best made after filtration, but the results will be unsatisfactory, showing a higher proportion of organic matter than will be found when the supply becomes clear.

The following methods of determining color and odor have been adopted by the Society of Public Analysts of Great Britain:—

Color.—A colorless glass tube, 2 feet long and 2 inches in diameter, is closed at each end with discs of colorless glass cemented on. An opening for filling and emptying the tube should be made at one end, either by cutting a small segment off the glass disc, or cutting out a small segmental section of the tube itself before the disc is cemented on. These tubes are most conveniently kept on hooks in a horizontal position, to prevent the entrance of dust.

The tube must be about half filled with the water to be examined, brought into a horizontal position, level with the eye, and directed toward a brightly illuminated white surface. The comparison of tint has to be made between the lower half of the tube containing the water under examination, and the upper half containing air only.

Odor.—Put about 150 c. c. of the water into a clean, wide mouth 250 c. c. stoppered bottle, which has been previously rinsed with the same water; insert the stopper and warm the water in a water bath to 100° F.; remove the bottle from the water bath, rinse it outside with good water, perfectly free from odor, and shake it rapidly for a few seconds; remove the stopper and immediately observe if the water has any smell. Insert the stopper and repeat the test.

In polluted waters the odors will sometimes give a clue to the origin of the pollution.

Reaction.—The determination of reaction is usually made by the addition of a neutral solution of litmus to the water. If an acid reaction is obtained the water should be boiled in order to determine if it is due to carbonic acid. Some of the more delicate indicators, such as phenolphthaleïn and lacmoid, may be used with advantage for these tests. The latter possesses the advantage that it is entirely unaffected by carbonic acid, but detects the slightest trace of free mineral acid. It is neutral, also, to normal metallic salts, such as ferrous sulphate, which are acid to litmus. Ferric salts, however, are acid to lacmoid. Its color changes are the same as those of litmus, $i.\ e.$, red with acids and blue with alkalies.

Phenolphthaleïn is best applied to the detection of weak acids, such as carbonic acid and the organic acids. In acid and neutral solutions it is colorless—in alkaline, deep red. Nearly all waters contain carbonic acid, and will therefore bleach a solution of phenolphthaleïn which has been reddened by a small amount of alkali.

TOTAL SOLIDS.

The following apparatus will be found convenient for this determination:—

A platinum basin holding 100 c.c. This should weigh about 45 grams; it should be kept clean and smooth by frequent burnishing with very fine white sea sand, a little of which should be placed in the palm of the hand, moistened, and the dish gently rubbed against it. The inside of the dish may be cleaned by fusing in it some potassium acid sulphate, washing with water and burnishing with sand. Very fine sea sand with round, smooth grains is the only kind suitable for this purpose. Coarse river sand, tripoli, or other rough polishing powders, must not be employed. If proper care is taken, the lustre of the dish will remain unimpaired indefinitely, and the loss in weight will be trifling. Neglect of this precaution will soon lead to serious damage to the dish. A small, smooth slab of iron or marble is convenient to set it on while cooling. When being heated over the naked flame the dish should rest on a triangle of iron wire, covered with pipe-stems.

A pair of platinum-pointed forceps, to be used in handling the dish. The platinum terminals may be kept bright and clean by the use of the sand.

The dishes of pure nickel have not been found by us to be satisfactory for this purpose.

The low-temperature burner used, as shown in the cut, will be found a very convenient substitute for the water-bath and hot air oven. When the burner is to be used for some hours, it will be necessary to wrap a wet rag around the rubber tube where it connects with the burner, to prevent the softening and consequent destruction of the tube by the heat.

The determination of total solids is made by evaporating

100 c.c. of the water in the platinum basin, which has been previously heated almost to redness, allowed to cool for ten minutes, and weighed. The operation is conducted over the low-temperature burner at a moderate heat, the dish being supported upon the crown. When the residue appears dry, the heat may be increased slightly for some minutes, or the dish may be removed to an air-oven, and kept at about 212° F. for a short time, then allowed to cool for ten minutes, and weighed. The above method will answer in most cases. In waters of exceptional purity it may be advisable to use larger quantities, such as 250 c.c. When the residue contains deliquescent bodies, the determination will not be accurate, and when bodies are present which take up much water of crystallization, the residue will need to be strongly heated, if control figures are to be obtained. This determination of total solids is described in connection with the technical examinations.

After the weight of the residue is obtained, the dish should be cautiously heated to low redness, and the effect noted. Nitrates and nitrites, calcium and magnesium carbonates, and magnesium chloride are decomposed; ammonium salts are driven off; potassium and sodium chlorides are also driven off if the temperature is high. Organic matter is at first charred, and by continued heating burned off. When the quantity of nitrates is considerable, slight deflagration may be observed, or the production of red fumes of nitrogen dioxide. The organic matter, in decomposing, not infrequently develops odors which indicate its character or source.

In water of high organic purity, the residue on heating will give no appreciable blackening nor odor, while in forest streams charged with vegetable matter derived from falling leaves, very decided blackening without unpleasant odor will be noticed. The loss of weight after heating cannot be taken as a measure of the organic matter present.

CHLORINE.

Solutions Required :—

Standard Silver Nitrate.—Dissolve about 5 grams of pure recrystallized silver nitrate in distilled water, and make the solution up to 1000 c. c. The amount of chlorine to which this is equivalent may be determined as follows : Several grams of pure sodium chloride are finely powdered and heated over a Bunsen burner for five minutes, not quite to redness. When cold, 0.824 gram is dissolved in water and the solution made up to 500 c. c. 25 c. c. of this solution should be treated as below, and the amount of silver solution required noted. Each c. c. of the sodium chloride solution is equivalent to .001 chlorine.

Potassium Chromate.—5 grams of potassium chromate are dissolved in 100 c.c. of distilled water. A solution of silver nitrate is added until a permanent red precipitate is produced, which is separated by filtration.

Analytical Process :—

If the preliminary test shows the chlorine to be present in considerable amount, the determination may be made on 100 c. c. of the water without concentration. If, however, there is but little present, 250 c. c. should be evaporated to about one-fifth, and the determination made on the concentrated liquid after cooling.

The water is placed in a porcelain dish or in a beaker standing on a white surface, a few drops of potassium chromate solution added, and standard silver nitrate solution run in from a burette until a faint red color of silver chromate remains permanent on stirring. The proportion of chlorine is then calculated from the number of c. c. of silver solution added. For greater accuracy a second determination may be made, using as a comparison the liquid first titrated, the red color having been previously discharged by a few drops of sodium chloride solution.

The water should always be as nearly neutral as possible before titration. If acid, it must be neutralized by the addition of some precipitated calcium carbonate.

NITROGEN IN AMMONIUM COMPOUNDS AND IN ORGANIC MATTER.

The nitrogen in ammonium compounds, and a part of that in the organic matter, is determined by a process of distillation first developed fully by Messrs. Wanklyn, Chapman and Smith. It depends upon the conversion of the nitrogen into NH_3 and its subsequent estimation in the distillate.

Apparatus Required:—

Distilling Apparatus.—That figured in the cut has been found to be the most convenient. The still consists of a Bohemian glass alembic of about 700 c. c. capacity, to which is attached, by means of a rubber stopper, a glass condenser. The heat is applied by means of the low-temperature burner, the iron ring of which is removed so that the alembic rests directly on the gauze. With this arrangement the heat is under perfect control, and the danger of fracturing the alembic is reduced to a minimum.

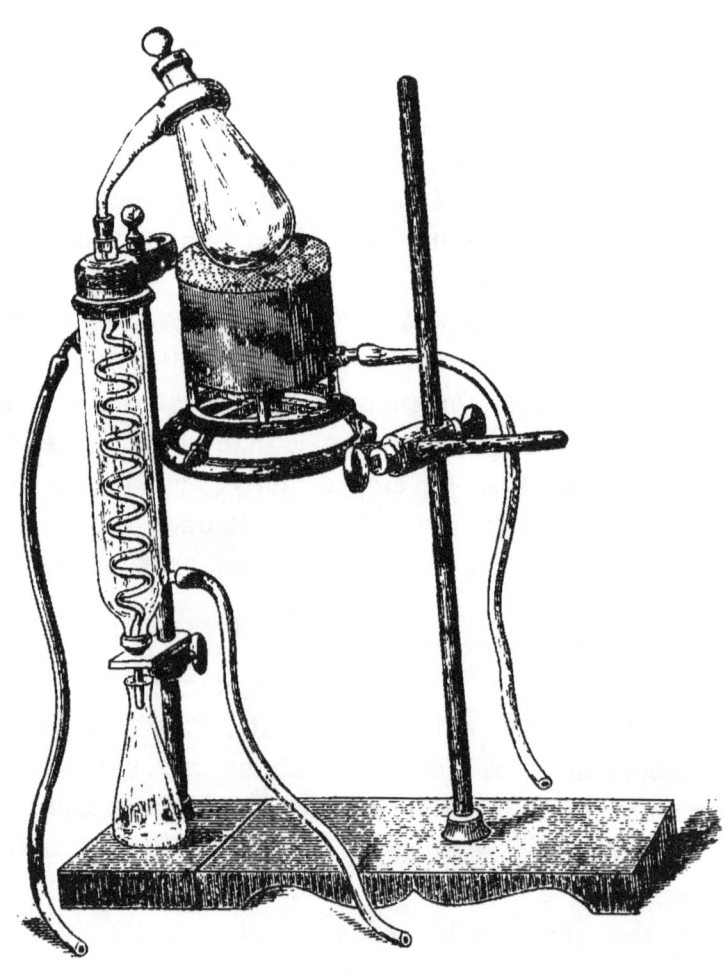

NITROGEN IN AMMONIUM COMPOUNDS.

Safety funnel, as in figure, for introducing liquids into the alembic. The bowl should hold about 50 c.c., and the inside diameter of the tube be about one-quarter inch. The lower end of the tube is protected by an outside tube, which renders it much easier to withdraw the funnel without soiling the neck of the alembic.

Cylinders for comparison-color tests, about one inch in diameter and holding 100 c.c., made of pure white glass. Otto Hehner has described an improved form, in which, by means of a stopcock at the base, the column of liquid can be drawn down at will. Two of this form will be needed.

Solutions Required:—

Sodium Carbonate.—A 20 per cent. solution of pure ignited sodium carbonate in ammonia-free water.

Ammonia-free Water.—If the distilled water of the laboratory gives a reaction with Nessler's test it should be treated with sodium carbonate, about one gram to the liter, and boiled until about one-fourth has been evaporated. Absolutely ammonia-free water may be obtained by distilling in a glass vessel, water made slightly acid with sulphuric acid.

Standard Ammonium Chloride.—Dissolve 0.382 gram of pure, dry, ammonium chloride in 100 c.c. of distilled water free from ammonia. For use, dilute 1 c.c. of this with pure distilled water to 100 c.c. 1 c.c. of this solution contains .00001 gram nitrogen.

Nessler's Reagent.—Dissolve 35 parts of potassium iodide in 100 parts of water. Dissolve 17 parts of mercuric chloride in 300 parts of water. The liquids may be heated to aid solution, but if so, must be cooled. Add the latter

C

solution to the former until a permanent precipitate is produced. Then dilute with a 20 per cent. solution of sodium hydrate to 1000 parts; add mercuric chloride solution until a permanent precipitate again forms; allow to stand until settled, and decant off the clear solution. The bulk should be kept in an accurately stoppered bottle, and a quantity transferred from time to time to a small bottle for use. The solution improves by keeping.

Alkaline Potassium Permanganate.—Dissolve 200 parts of pure potassium hydroxide and 8 parts of pure potassium permanganate in 1000 parts of distilled water.

Analytical Process:—

The alembic and condenser are thoroughly rinsed out with ammonia-free water, 500 c. c. of the water to be tested introduced, about 5 c. c. of the sodium carbonate solution added to render the water alkaline, and a piece of pumice stone heated to redness and dropped in while hot. The water is then boiled gently until the distillate measures 50 c. c. The distillate is transferred to one of the color comparison cylinders and 2 c. c. of the Nessler's reagent added. A yellow color is produced, the intensity of which is proportional to the amount of NH_3 present. The full color is developed in from three to five minutes. This color is exactly matched by introducing into another cylinder 50 c. c. of ammonia-free water, some of the standard ammonium chloride solution, and 2 c. c. Nessler's reagent, as before. According as the color so produced is deeper or lighter than that obtained from the water, other comparison liquids are prepared containing smaller or larger proportions of the ammonium chloride, until the proper color is produced.

The above process may be much shortened by the use of

Hehner's cylinders. The distillate is placed in one of these and the Nessler's reagent added. In another, the comparison liquid is prepared, using a constant quantity, say one c. c., of the ammonium chloride solution, or, what is better, an amount which is thought to be about what is present in the first cylinder. When the colors have fully developed, the darker liquid is run down by means of the stopcock until the two tints agree. The determination is made more exact, if the volumes of the two liquids are equalized by the addition of water to the smaller.

The distillation is continued, and successive portions of 50 c. c. of the distillate taken and tested, until the liquid no longer reacts with Nessler's reagent. The sum of the figures obtained for the several distillates gives the total nitrogen existing in ammonium compounds.

The residue in the alembic serves for the determination of that portion of the nitrogen of the organic matter which is converted by an alkaline solution of potassium permanganate into ammonia—the so-called "albuminoid ammonia" of Messrs. Wanklyn, Chapman and Smith.

50 c. c. of the alkaline permanganate solution is placed in a porcelain dish of about 150 c. c. capacity, the dish nearly filled with distilled water and the liquid boiled down to about its original volume—50 c. c. This is added to the alembic by means of the safety funnel, the distillation resumed, and the ammonia estimated in each 50 c. c. of the distillate as before.

It is the practice of some analysts to mix the distillates of each of the above operations, and thus make determinations merely of the total quantity of N in ammonium compounds and the total N "by permanganate." By so doing valuable information may be lost, since it has been pointed out by several observers, notably Mallet, that the ammonia may be differently distributed in the distillates according to the state, decomposing or otherwise, in which the organic matter exists in the water. Certain other classes of water yield their ammonia in a characteristic manner, which will be referred to further on. Mallet has further pointed out that many waters may contain substitution ammoniums which may pass over before the addition of the alkaline permanganate, but not be correctly measured by Nesslerizing. To avoid this source of error, he has suggested that two determinations be made on each sample, one as above described and the other by the addition of alkaline permanganate without previous distillation of the water. In this manner a higher figure will often be obtained than the sum of the figures from the two distillations by the other process.

Since small quantities of ammonium compounds and nitrogenous matters are everywhere present, the greatest care should be exercised in order to avoid their introduction in any way during the course of the analysis. All measuring vessels, cylinders, etc., should be thoroughly rinsed before using.

NITROGEN AS NITRATES.

Solutions Required:—

Acid Phenyl Sulphate.—18.5 c.c. of strong sulphuric acid are added to 1.5 c.c. of water and 3 grams of pure phenol. Preserve in a tightly-stoppered bottle.

Standard Potassium Nitrate.—0.722 gram of potassium nitra'e, previously heated to a temperature just sufficient to fuse it, is dissolved in water, and the solution made up to 1000 c.c. 1 c.c. of this solution will contain .0001 grm. of nitrogen.

Analytical Process :—

A measured volume of the water is evaporated just to dryness in a platinum or porcelain basin. 1 c.c. of the acid phenyl sulphate is added and thoroughly mixed with the residue by means of a glass rod. 1 c.c. of water is added, three drops of strong sulphuric acid, and the dish gently warmed. The liquid is then diluted with about 25 c.c. of water, ammonium hydroxide added in excess, and the solution made up to 100 c.c.

The reactions are :—

Acid phenyl sulphate.   Trinitrophenol (picric acid).
$$HC_6H_5SO_4 + 3\,HNO_3 = HC_6H_2(NO_2)_3O + H_2SO_4 + 2H_2O.$$

Ammonium picrate.
$$HC_6H_2(NO_2)_3O + NH_4HO = NH_4C_6H_2(NO_2)_3O + H_2O.$$

The ammonium picrate imparts to the solution a yellow color, the intensity of which is proportional to the amount present.

Five c.c. of the standard solution of potassium nitrate is now similarly evaporated in a platinum basin treated as above, and made up to 100 c.c. The color produced is compared to that given by the water; and one or the other of the solutions diluted until the tints of the two agree. The comparative volumes of the liquids furnish the necessary data for determining the amount of nitrate present, as the following example will show :—

Five c.c. of standard nitrate is treated as above, and made up to 100 c.c.

.0001
   5
.0005 gram N per 100 c.c.
  10
.0050   "   "   1000 "

Suppose 100 c.c. water similarly treated is found to require dilution to 150 c.c. before the tint will match that of the standard; then

$$100 : 150 :: .005 : .0075$$

i. e., water contains 7.5 milligrams of nitrogen as NO_3 per liter.

The ammonium picrate solution keeps very well, especially in the dark. A good plan, therefore, is to make up a standard solution equivalent to, say, ten milligrams of nitrogen as nitrate per liter, to which the color obtained from the water may be directly compared.

The results obtained by this method are quite accurate. Care should be taken that the same quantity of acid phenyl sulphate is used for the water and for the comparison liquid, otherwise different tints instead of depths of tints are produced.

With subsoil and other waters probably containing much nitrates, 10 c.c. of the sample will be sufficient for the test, but with river and spring waters, 25 to 100 c.c. may be used. When the organic matter is sufficient to color the residue, it will be well to purify the water by addition of alum and subsequent filtration, before evaporating.

NITROGEN AS NITRITES.

Solutions Required :—

Naphthylammonium Chloride.—Saturated solution in water free from nitrites. It should be colorless; a small quantity of animal charcoal allowed to remain in the bottle will keep it in this condition.

Para-amido-benzene-sulphonic Acid [*Sulphanilic Acid*].— Saturated solution in water free from nitrites.

NITROGEN AS NITRITES.   31

Hydrochloric Acid.—25 c.c. of concentrated pure hydrochloric acid added to 75 c.c. water free from nitrites.

Standard Sodium Nitrite.—0.275 gram pure silver nitrite is dissolved in pure water, and a dilute solution of pure sodium chloride added until the precipitate ceases to form. It is then diluted with pure water to 250 c.c., and allowed to stand until clear. For use 10 c.c. of this solution are diluted to 100 c.c. It is to be kept in the dark.

One c.c. of the dilute solution is equivalent to .00001 gram nitrogen.

The silver nitrite is prepared thus: A hot concentrated solution of silver nitrate is added to a concentrated solution of the purest sodium or potassium nitrite available, filtered while hot and allowed to cool. The silver nitrite will separate in fine needle-like crystals, which are freed from the mother liquor by filtration by the aid of a filter pump. The crystals are dissolved in the smallest possible quantity of hot water, allowed to cool and again separated by means of the pump. They are then thoroughly dried in the water bath, and preserved in a tightly-stoppered bottle away from the light. Their purity may be tested by heating a weighed quantity to redness in a tared porcelain crucible and noting the weight of the metallic silver. 154 parts of $AgNO_2$ leave a residue of 108 parts Ag.

Analytical Process :—

100 c.c. of the water is placed in one of the color-comparison cylinders, the measuring vessel and cylinder having previously been rinsed with the water to be tested. By means of a pipette, one c.c. each of the solutions of sulphanilic acid, dilute hydrochloric acid and naphthylammonium chloride is dropped into the water in the order named. It is convenient to have three pipettes for this test, and to use

them for no other purpose. In any case the pipette must be rinsed out thoroughly with nitrite-free water each time before using, as nitrites in quantity sufficient to give a distinct reaction may be taken up from the air.

One c.c. of the standard nitrite solution is placed in another clean cylinder, made up with nitrite-free water to 100 c.c. and treated with the reagents as above.

In the presence of nitrites a pink color is produced, which in dilute solutions may require half an hour for complete development. At the end of this time the two solutions are compared, the colors equalized by diluting the darker, and the calculation made as explained under the estimation of nitrates.

The following are the reactions:—

Para-amido-benzene-sulphonic acid.   Nitrous acid.   Para-diazo-benzene-sulphonic acid.
$$C_6H_4NH_2HSO_3 + HNO_2 = C_6H_4N_2SO_3 + 2H_2O$$

Naphthylammonium chloride.   Azo-alpha-amido-naphthalene-parazo-benzene-sulphonic acid.
$$C_6H_4N_2SO_3 + C_{10}H_7NH_3Cl = C_{10}H_6(NH_2)NNC_6H_4HSO_3 + HCl$$

The last-named body gives the color to the liquid.

OXYGEN-CONSUMING POWER.

All organic materials being more or less easily oxidized, several methods have been suggested for determining the oxygen-consuming powers of waters by treatment with active oxidizing agents. These methods have, however, proved to be limited in value. The organic matters in water are very variable in character and condition, and their oxidability is subject to much difference, according to the circumstances under which the test is made. Nevertheless, as a high oxygen-consuming power certainly indicates departure from purity, some additional evidence may be obtained by the use of these methods. Potassium permanganate is especially suitable for this purpose, and the test is usually made by introducing a definite amount of it into the water, rendered slightly acid, and measuring after a definite period the proportion of the permanganate which has been decomposed.

The following is suggested as a convenient method of approximating the oxygen-consuming power of a water :—

Solutions Required :—

Standard Permanganate.—.395 gram pure potassium permanganate is dissolved in distilled water, and the solution made up to 1000 c.c. 1 c.c. is equal to .0001 gram oxygen.

Diluted Sulphuric Acid.—Add 50 c.c. of pure sulphuric acid to 100 c.c. of water, and then add solution of potassium permanganate until a faint pink color is obtained, which is permanent when the liquid is heated to 80° Fahrenheit for four hours.

Analytical Process :—

Five stoppered bottles holding 500 c.c. are cleaned successively by the use of about 25 c.c. of strong sulphuric acid, the acid being brought in contact with every portion of the interior, then thoroughly rinsed with clean water, and 250 c.c. of the water to be tested put in each one ; 10 c.c. of the dilute sulphuric acid is added to each, together with regularly increasing quantities of the standard permanganate solution, say 2, 4, 6, 8 and 10 c.c., respectively. At the end of an hour they should be examined, and a note made as to which bottles, if any, are decolorized. They should be again examined at the end of three more hours, and again at the expiration of twenty-four hours. If at the expiration of the first, or even the fourth hour, all the bottles are decolorized, an additional 10 c.c. of the permanganate solu-

D

tion should be added to each one, allowed to stand, and the effect noted as before.

With ordinary waters it will be found that the first and probably the second bottle will be decolorized, while some color will remain in the third, and it will be but little diminished in the fourth and fifth. In this way an outside figure for the oxygen-consuming power may be obtained, which is all that is essential. If a closer figure is desired, the experiment can be repeated, using quantities of permanganate intermediate between those marking the limits of the reaction; for instance, if it is found that the second bottle is completely decolorized, while the third bottle still retains a trace of tint, another sample may be prepared, using 5 c.c. of the permanganate, and thus the determination of the effect may be brought within the limits of one milligram of oxygen.

The advantage of the above method of procedure is the ease with which the rate of oxidation may be observed. This is considered by some to be of more importance than the actual amount of oxygen consumed.

It must not be lost sight of that if a water contains nitrites, ferrous compounds, or sulphur compounds other than sulphates, the proportion of oxygen consumed will be greater than that required for the organic matter. If hydrogen sulphide is present, the water should be boiled and cooled before making the determination. It has been proposed, in order to remove the nitrites before applying the permanganate, to take 500 c.c. of the water, add 10 c.c. of the dilute sulphuric acid, boil for twenty minutes, allow to cool, and then treat with permanganate. Since, however, the amount of nitrites, if appreciable, can be directly determined, it is more satisfactory to deduct from the oxygen consumed the amount of O required to convert the nitrites present into nitrates, and the remainder will be the amount of O required for the other oxidizable ingredients. 14 parts of N existing as nitrite require 16 parts of O to oxidize to nitrate. Similarly, 112 parts of iron in a ferrous compound will require 16 parts of oxygen to oxidize to the ferric condition.

More exact determinations of the oxygen consumed may be made by the following method, which is due in the main to Dr. Tidy, has been improved by Dr. Dupré, and is approved by the Society of Public Analysts of Great Britain :—

Solutions Required :—

Standard potassium permanganate and *diluted sulphuric acid* described above.

Potassium Iodide.—10 grams of the pure salt recrystallized from alcohol, dissolved in 100 c. c. of distilled water.

Sodium Thiosulphate.—1 gram of the pure crystallized salt dissolved in 1000 c. c. of distilled water.

Starch Indicator.—1 gram of clean starch is mixed smoothly with cold water into a thin paste, then poured gradually into about 200 c. c. of boiling water, the boiling continued for one minute, allowed to settle, and the clear liquid used. It is best made up fresh each time it is required.

Analytical Process :—

Two determinations are made, one, of the oxygen consumed in fifteen minutes, which is considered to represent the nitrites, sulphides or ferrous compounds, and the other of the oxygen consumed by four hours' action. Both determinations are made at a temperature of 80° F. Three glass-stoppered bottles, of about 350 c. c. capacity, are rinsed with strong sulphuric acid, and then with water. In one is placed 250 c. c. of pure distilled water as a control experiment, and each of the others 250 c. c. of the water to be tested. The bottles are stoppered, and brought to a temperature of 80° F. ; 10 c. c. of the dilute sulphuric acid and .10 c. c. of the standard permanganate are added to each and the stoppers again replaced. At the end of fifteen minutes one sample of water is removed from the bath, and two or three drops of the potassium iodide solution added to remove the pink color. After thorough admixture the thiosulphate solution is run in from a burette until the yellow color is nearly destroyed, a few drops of the starch solution added, and the addition of the thiosulphate continued until the blue color is quite discharged. If the addition of the thiosulphate solution has been properly conducted, one drop of the permanganate solution will restore the blue color.

The other bottles are maintained at 80° F. for four hours. Should the pink color disappear rapidly in the bottle containing the water under examination, 10 c. c. of the permanganate solution must be added to each bottle, in order to maintain a distinct pink color. At the end of four hours each bottle is removed from the bath, two or three drops of potassium iodide added, and the titration with thiosulphate solution conducted as just described. The calculation is most conveniently made as follows:—

$a =$ number of c. c. required for the control experiment.
$b =$ number of c. c. required for the water under examination.
$c =$ available O in permanganate employed (.001 when 10 c. c. are employed).
$x =$ oxygen consumed by water.

Then, $a : a-b :: c : x$.

The oxygen-consuming power may also be indirectly estimated by the action of the organic matter upon silver compounds. H. Fleck's method (*Fresenius Quantitative Analysis, English edition, vol. ii, p. 127*) depends upon the reduction produced by boiling the water with alkaline solution of silver thiosulphate and estimation of the unreduced silver. A. R. Leeds (*Lond., Edin. and Dub. Phil. Mag., July, 1883*) gives a method by treating the water with decinormal silver nitrate, exposing to light until it settles perfectly clear, and estimating the reduced silver in the deposit.

These methods are open to practically the same objections as in the use of permanganate, and do not seem to possess any decided advantage. Qualitative results of some interest may occasionally be obtained by the following method: 2 c.c. of a one per cent. solution of silver nitrate, rendered decidedly alkaline by ammonium hydroxide, are added to 100 c.c. of the water in a stoppered bottle, which is then placed in full sunlight for two hours. Waters containing but little organic matter will not show at the end of this period any appreciable tint. The following results will show the character of the test:—

```
Schuylkill water, . . . . . . . . . . . . . no color.
    "       " with 0.02 c.c. urine, . . . red-brown.
    "       " with 0.01 c.c. urine, . . . deep brown.
    "       " with 0.25 gram raw sugar, . no color.
Well water, not perfectly pure, but
            not unfit to drink, . . . . . faint black.
    "    "   markedly contaminated, . .  black ppt. almost
                                          immediately.
Water from a small stream, quite pure, . . . no color.
```

PHOSPHATES.

Solutions Required:—

Ammonium Molybdate.—Ten grams of molybdic anhydride are dissolved in a mixture of 15 c.c. of concentrated ammonia, sp. gr. .90, and 25 c.c. water. This solution is poured slowly, and with constant stirring, into a mixture of 65 c.c. of concentrated nitric acid, sp. gr. 1.4 and 65 c.c. water, and allowed to stand until clear. It keeps best in the dark.

Analytical Process :—

One liter is evaporated to about 50 c.c. A few drops of a dilute solution of ferric chloride added and ammonia in slight excess. The precipitate of ferric hydroxide, which contains all the phosphate, is filtered off and dissolved on the filter by the smallest possible quantity of hot dilute nitric acid. The filtrate and washings should not exceed 5 c.c.; if more, it should be evaporated to this bulk. The liquid is heated nearly to boiling, 2 c.c. of ammonium molybdate solution added, and the liquid kept at a temperature of about 60° F. for half an hour. If the quantity of precipitate is appreciable, it is collected on a small weighed filter, washed with distilled water, dried at 100° F. and weighed. The weight of the precipitate multiplied by 0.05 gives the amount of PO_4. If the quantity is not sufficient to collect in this manner, it is usually reported, according to circumstances; as "traces," "heavy traces," or "very heavy traces."

SUGAR-TEST.

This method was proposed by Heisch for the detection of a fungus supposed to be peculiar to sewage. It consists in adding to the water a small quantity of sugar in which the fungus grows with great rapidity. The test is applied as follows :—

SUGAR-TEST.

A stoppered bottle of about 100 c.c. capacity is rinsed thoroughly with the water to be tested, filled with the clear water, about half a gram of pure crystallized sugar added and the stopper inserted. The bottle is placed in a strong light, and kept at a temperature of about 80° F. At the end of several hours it is examined for the fungus, which, in a good side light, with the bottle against a dark background, is generally easily visible as a distinct turbidity. If examined under a power of 250 diameters, it is found to consist of small spherical cells, having in most cases a bright nucleus. After the lapse of some days these gradually group themselves together something like grapes; they next spread out into strings, with a surrounding wall connecting the cells together; the original cells then seem to break and leave apparently tubular threads joined by branches. In the more marked cases the development of the fungus is accompanied by a distinct odor of butyric acid. Hydrogen sulphide and other gases are also sometimes formed. These odors do not generally appear until after twenty-four hours or longer. Heisch concluded that the cells thus developed were distinct evidence of sewage contamination, but Frankland has shown that the germs are present in all waters that have been exposed even momentarily to the air, and that their development is due simply to the presence of phosphates in the water. He found that the addition even of minute traces of a phosphate, either as sodium phosphate, white of egg, or animal charcoal, at once determined the growth in saccharine waters which, before, exhibited no tendency to develop them.

DISSOLVED OXYGEN.

The method here given, a modification of Mohr's, was recently proposed by Blarez. Our experiments indicate that it is rapid and satisfactory.

Solutions Required:—

Sodium Hydroxide.—40 grams of pure sodium hydroxide to the liter.

Ferrous-Ammonium Sulphate.—40 grams dissolved in about a liter of water, and acidified with a few drops of concentrated sulphuric acid.

Decinormal Potassium Permanganate.—3.156 grams dissolved in a liter of distilled water. The accuracy of this solution should be determined by titration with a known weight of ferrous-ammonium sulphate. One c.c. should be equivalent to .0008 gram of oxygen.

The apparatus employed (shown in the annexed cut) is a globular separator, of about 250 c.c. capacity. Above the bulb is a caoutchouc stopper carrying a cylindrical funnel, of about 12 c.c. capacity, terminating in a tube, $\frac{1}{6}$ inch calibre, sharply contracted at the outlet to a capillary opening. The tube should project about $\frac{1}{4}$ inch below the stopper. The exact capacity of the apparatus is measured as follows: The bulb is completely filled with water and the stopper inserted; the level of the water will rise slightly in the funnel tube, and should be brought down to its outlet by drawing a little off at the stopcock, after which the water is run into a graduated measure and its volume noted.

Analytical Process :—

35 c.c. of mercury and 10 c.c. of sodium hydroxide solution are put into the bulb, and then sufficient of the water to be tested to fill it. The funnel stopper is inserted and the water which rises into the funnel brought into the bulb by cautiously running out at the stopcock, mercury, the volume of which should be noted. The exact volume of water used is thus known. Five c.c. of the ferrous-ammonium sulphate solution are poured into the funnel, brought into the bulb by running out mercury, and the liquid thoroughly mixed by giving the apparatus a gyratory movement. After standing five or six minutes the oxygen will be completely absorbed; 10 c.c. of the diluted sulphuric acid are now added by the same method. On agitating the bulb the contents become clear. The watery liquid is then transferred to a beaker and titrated with decinormal permanganate. A volume of water equal to that used in the test is poured into another beaker, 10 c.c. each of the sodium hydroxide and diluted sulphuric acid added, and then 5 c.c. of ferrous-ammonium sulphate solution. The resulting liquid is titrated with permanganate. The weight of oxygen corresponding to the difference between the two titrations gives the weight of dissolved oxygen in the liquid employed. From this should be subtracted as correction the amount of oxygen introduced with the sodium hydroxide used (that in the sulphuric acid has no appreciable effect). This is found by reference to the table.

Nitrates do not appear to impair the accuracy of this method, and the interfering action of nitrites and other reducing compounds is avoided by the control experiment as detailed.

It is perhaps hardly necessary to add that the exact

temperature of the water is to be noted at the time of collection of the sample.

In transferring to the bulb the water should be agitated as little as possible in contact with the air, in order to avoid the absorption of oxygen. A syphon should be used for this purpose, the lower end being allowed to reach to the bottom of the bulb. It is well, also, to fill the bulb with carbon dioxide before allowing the water to enter.

The following modification is suggested as being especially suitable for poorly oxygenated waters: An accurately stoppered bottle, the exact capacity of which is known (about 500 c.c. is a convenient size), is completely filled at the source with the water to be examined, and the stopper inserted so as to drive out all air. The stopper is removed in the laboratory, 50 c.c. of the water drawn off with a pipette, and the water covered immediately with a layer of gasoline previously purified by shaking up several times with a solution of potassium permanganate and diluted sulphuric acid, and washed several times with water. The sodium hydroxide, ferrous-ammonium sulphate and sulphuric acid are introduced into the water by means of burettes to which long glass delivery tubes are attached. The titration with potassium permanganate is conducted in the same way. The liquid is mixed from time to time, as the solutions are added, by means of a glass rod. In this way the air may be completely excluded throughout the entire operation. The amount of water titrated is, of course, equal to the whole capacity of the bottle, less the 50 c.c. removed by the pipette.

The control experiment on an equal volume of the water, and the correction for the oxygen added with the sodium hydroxide solution, are made as detailed above.

PYROGALLOL METHOD.

An approximate determination of the amount of oxygen dissolved in a water may be rapidly made as follows:—

Solutions Required :—

Pyrogallol.—25 grams of pure pyrogallol dissolved in 100 c. c. distilled water.

Sodium Hydroxide.—5 grams of pure sodium hydroxide dissolved in 100 c.c. distilled water.

Caramel Solution.—This is made by heating sugar in a porcelain dish or crucible to about 400° F. for a short time and dissolving the resulting mass in water. It is standardized as follows: A bottle holding about 250 c. c. is completely filled with thoroughly aërated water and 5 c. c. each of the pyrogallol and sodium hydroxide solutions added by means of a long, graduated pipette, which is inserted about half-way into the bottle. As the solution runs in, the pipette is gradually raised so that when it is completely withdrawn the liquid fills the bottle completely. The stopper is now inserted and the liquid mixed by rotating the bottle. The solution takes on a reddish-brown color, the intensity of which is dependent on the amount of free oxygen present. A bottle of the same size is now nearly filled with clear water and the caramel solution run in from a burette until the color matches that in the first bottle. The number of c. c. required is noted. The exact quantity of oxygen contained in the thoroughly aërated water being ascertained from the table, the oxygen equivalent to each c. c. of the caramel solution is readily ascertained.

Analytical Process :—

A bottle completely filled with the water to be tested is treated with sodium hydroxide and pyrogallol as described above, the stopper inserted, and the tint so produced exactly matched by nearly filling another bottle of the same size with clear water and running in caramel solution from a burette. From the number of c. c. of caramel solution required, the amount of free oxygen in the water is calculated. Compressed pellets of pure pyrogallol of definite weight may be used instead of the solution mentioned above. This method may be suitable for rapid examination of waters in sanitary surveys.

Dupré has employed the determination of free oxygen for the estimation of the proportion of oxygen-consuming microbes. The principle of the method is that pure water, if kept in a closed bottle, will neither gain nor lose oxygen in any length of time, but if organisms capable of causing absorption of oxygen are present, the quantity will decrease.

The experiment is carried out by placing a sample of the water in a clean bottle, and vigorously shaking it to saturate with air. A clean 250 c.c. bottle is completely filled with the water, tightly stoppered, and maintained at a temperature of 68° F. for ten days; the free oxygen remaining is then determined.

POISONOUS METALS.

In this class are included *barium, chromium, zinc, arsenic, copper* and *lead; manganese* and *iron* also, though not usually classed in this group, are objectionable when present in notable amounts.

Barium is rarely present, and only in water containing no sulphates. It can be detected and estimated by slightly acidifying the water with hydrochloric acid, filtering if necessary, and adding solution of calcium sulphate. The precipitated barium sulphate is collected and weighed in the usual way.

Chromium is rarely present, but may be looked for in

the waste waters of dye works and similar sources. To detect it, a considerable volume of the water is evaporated to dryness with addition of a small amount of potassium chlorate and nitrate, transferred to a porcelain crucible and brought to quiet fusion; any chromium present will be found in the residue in the form of chromate. The fused mass, after cooling, is boiled with a little water, filtered, the filtrate rendered slightly acid with hydrochloric acid, and a solution of hydrogen dioxide added. In the presence of chromium a transient blue color will appear; by adding a little ether, and shaking the mixture the color will pass into the ether, and on standing will form a blue layer on the surface of the water.

Zinc is best detected by the test described by Allen. The water is rendered slightly ammoniacal, heated to boiling, filtered, and the clear liquid treated with a few drops of potassium ferrocyanide; in the presence even of the merest trace of zinc a white precipitate will be produced.

Arsenic may be detected by Marsh's test, as follows: A liter of the water is made alkaline by the addition of a little sodium hydroxide, evaporated to very small bulk, treated with hydrochloric acid, and transferred to a hydrogen-generating apparatus. This consists of a small bottle containing a few pieces of zinc covered with water; through the stopper are passed a thistle tube and a delivery tube bent at right angles; to the latter is attached a small tube containing calcium chloride, to dry the gas. Before the addition of the water residue, sulphuric acid is added through the funnel tube, and, after sufficient time has elapsed for the air in the apparatus to be completely expelled by the hydrogen, a light is applied and a porcelain crucible cover held in the flame. If the reagents used are free from arsenic, no deposit will be formed on the porcelain. The concentrated water is now added by the funnel tube; if it contains arsenic, steel-gray spots with metallic lustre will be obtained by holding the porcelain cover in the flame. A point in the delivery tube beyond the calcium chloride should also be heated by the Bunsen burner; a dark ring of arsenic will be formed in the tube just beyond the point of application of the heat. The latter test is the more delicate.

Iron is detected by the addition of a drop of ammonium sulphide to the water in a tall glass cylinder. Ferrous sulphide is formed, having a greenish-black color, instantly discharged by acidifying the water with dilute hydrochloric acid. A still better test is the production, with a solution of potassium sulphocyanate, of a blood-red color, due to the formation of ferric sulphocyanate. The water should be first boiled with a few drops of nitric acid, to convert the iron to the ferric condition, cooled, and a drop or two of the solution of potassium sulphocyanate added. The test is very delicate. Either of the above tests may be made quantitative by matching the color produced in 100 c.c. of the water with that obtained from a known weight of iron. The method with potassium sulphocyanate is preferable, as it is more delicate and there are fewer interfering conditions. The following is the method as elaborated by Thompson and described in Sutton's "Volumetric Analysis:"—

SOLUTIONS REQUIRED :—

Standard Ferric Sulphate.—0.7 gram ferrous ammonium sulphate is dissolved in water acidified with sulphuric acid, and potassium permanganate solution added until the solution turns a very faint pink color. The solution is diluted to a liter. 1 c. c. contains 0.1 milligram iron.

Diluted Nitric Acid.—30 c. c. concentrated nitric acid diluted with water to about 100 c. c.

Potassium Sulphocyanate.—5 grams of the salt dissolved in about 100 c. c. water.

ANALYTICAL PROCESS:

About 100 c. c. of the water is evaporated to small bulk, acidified with hydrochloric acid, and just sufficient dilute potassium permanganate solution added to convert all the iron to the ferric condition. The liquid is evaporated nearly to dryness to drive off excess of acid, then diluted to its original volume, 100 c. c. Into two tall glasses marked at 100 c. c., 5 c. c. of the nitric acid and fifteen c. c. of the sulphocyanate solution are added. To one of these a measured volume of the water treated is added and both vessels filled up to the mark with distilled water. If iron is present, a blood-red color will be produced. Standard iron solution is added to the second vessel until the color agrees. The amount of water which is added to the first glass will depend upon the quantity of iron it contains; not more should be used than will require two or three c.c. of the standard to match it, otherwise the color will be too deep for comparison.

Manganese.—The following method is described by Wanklyn in his treatise on water analysis. About one liter of the water is evaporated to small bulk, nearly neutralized by hydrochloric acid and treated with a few drops of a solution of hydrogen dioxide. The formation of a brown precipitate indicates the presence of manganese. The test is very delicate. The precipitate may be collected on a filter, the filter ashed, and the residue fused with a mixture of sodium carbonate and potassium nitrate. Green potassium manganate will be produced, which, when boiled with water, will give a bright red solution of potassium permanganate. The quantitative determination is given elsewhere.

Lead may be readily detected by adding to the water in a tall glass cylinder a drop of ammonium sulphide; brownish black lead sulphide is formed, which does not dissolve either by acidulating the water with dilute hydrochloric acid (distinction from iron) nor by the addition of about one c.c. of a strong solution of potassium cyanide (distinction from copper). The result may be confirmed by the production of lead chromate by addition of potassium chromate. Allen describes a method of manipulating this test which is very satisfactory. A drop of potassium chromate solution is cautiously added to the water in a tall vessel, so that it sinks slowly through the liquid. In this way a very faint cloud of lead chromate may be readily observed. The liquid must be viewed with a dark background. The addition of acetic acid diminishes the delicacy of the test.

In the absence of copper the amount of lead present may be determined as follows: A solution is prepared containing 1.6 grams of lead nitrate to the liter; one c.c. of this contains one milligram lead. 100 c.c. of the water to be tested is placed in a tall glass vessel, made acid by the addition of a few drops of acetic acid and five c.c. of hydrogen sulphide added. In a similar vessel 100 c.c. of distilled water is placed, together with the same quantities of acetic acid and hydrogen sulphide, and sufficient of the standard lead solution to match the tint in the first cylinder. The amount of lead in the water under examination is thus known.

Copper is detected in the same manner as lead by acidifying the water with acetic acid and adding hydrogen sulphide water. The precipitate is distinguished from lead sulphide by the fact that the color is discharged on the addition of about one c.c. of a strong solution of pure potassium cyanide. It may be further confirmed by the addition to another portion of the water of a solution of potassium ferrocyanide. In the presence of even a very small amount of copper a mahogany red color is produced.

In the absence of lead, copper is estimated in the same way as that metal, using, however, a standard solution of copper for the comparison liquid. This is made by dissolving 3.929 grams of crystallized copper sulphate in one liter of water. One c.c. of the solution contains one milligram copper.

If both lead and copper are present, a large quantity of the water should be evaporated to small bulk, and the metals separated and estimated by any one of the ordinary laboratory methods.

The following table, prepared by A. J. Cooper (*Jour. Soc. Chem. Ind.*, Feb., 1886) indicates the comparative delicacy of some of the ordinary tests for the detection of poisonous metals in water:—

Metal.	Reagent.	Depth of Liquid, 3¾ inches.	Depth of Liquid, 14½ inches. Cylinder enclosed in opaque tube.
Copper...	K_4Cy_6Fe	1 part of metal detected in 4,000,000 of water.	1 part of metal detected in 11,750,000 of water.
" ...	NH_4HO	1,000,000 "	1,950,000 "
" ...	H_2S	4,150,000 "	15,660,000 "
Zinc....	NH_4HS	2,500,000 "	...
Arsenic...	H_2S	3,600,000 "	7,520,000 "
Lead.....	K_2CrO_4	4,000,000 "	5,875,000 "
" ...	H_2S	100,000,000 "	196,000,000 "

TECHNICAL EXAMINATIONS.

GENERAL QUANTITATIVE ANALYSIS.

Silica, Iron, Aluminum, Manganese, Calcium, and Magnesium.—One liter of the water acidified with hydrochloric acid is evaporated to complete dryness, best in a platinum dish, the residue treated with hydrochloric acid and water, and the separated *silica* filtered, washed, dried, ignited in a platinum crucible and weighed.

To the filtrate previously boiled with a few drops of strong nitric acid a slight excess of ammonia is added, the liquid boiled several minutes, the precipitate collected upon a filter, washed thoroughly with boiling water, dried, ignited and weighed. It consists of Fe_2O_3 and Al_2O_3. It is usual to report their combined weight without separation. It also contains all the phosphates and some manganese if much is present in the water. In such cases the precipitate after collection on the filter is re-dissolved in hydrochloric acid and neutralized with a dilute solution of ammonium carbonate until the water almost becomes turbid. It is then boiled and the precipitate, now free from manganese, washed, dried, ignited and weighed.

If no manganese or only traces are present, the filtrate from the iron is mixed with sufficient ammonium chloride to prevent the precipitation of the magnesium, ammonium hydroxide, and then ammonium oxalate added in quantity sufficient to precipitate the calcium and to convert all the magnesium into oxalate, and thus hold it in solution. The precipitate contains all the calcium and a small quantity of the magnesium. If the magnesium is present only in small quantity as compared with the calcium, the amount carried down is small and may be disregarded;

otherwise a second precipitation should be made as follows: The solution is allowed to stand until the precipitate has subsided; this will require some hours. The supernatant liquid is poured off through a filter, the precipitate washed once in the same way by decantation, then dissolved in hydrochloric acid, water added, and then ammonia and a small quantity of ammonium oxalate. After the precipitate of calcium oxalate has thoroughly subsided it is filtered off, washed, dried, placed with the filter in a weighed platinum crucible, ignited over the Bunsen burner for a short time, and then over the blast lamp for from five to fifteen minutes. The calcium is thus obtained in the form of oxide, which is allowed to cool in the desiccator and weighed. The weight thus obtained multiplied by 0.7142 gives the weight of *calcium*.

The filtrates are mixed, slightly acidified with hydrochloric acid, concentrated and cooled, sodium phosphate and ammonium hydroxide added in excess, mixed, and allowed to stand in the cold for about twelve hours. The precipitated ammonium magnesium phosphate is brought upon a filter, that adhering to the sides of the vessel being dislodged by rubbing with a glass rod tipped with a piece of clean rubber tubing. It is washed with a solution of ammonium hydroxide, made by mixing one part of the ammonia of 0.96 sp. gr. with three parts water. The precipitate is then dried, transferred to a platinum crucible, the filter ashed separately and added to it, and the whole heated at first gently and then to intense redness for several minutes. After cooling it is weighed. It consists of magnesium pyrophosphate; the weight multiplied by 0.2162 gives the weight of *magnesium*.

Manganese, if present in appreciable quantity, is separated before the precipitation of the calcium, as follows: The filtrate from the iron precipitate is slightly acidulated with hydrochloric acid, concentrated, and the manganese precipitated as sulphide by colorless or slightly yellow solution of ammonium sulphide. The flask, which should be nearly full, is stoppered, allowed to rest in a moderately warm place until the precipitate has thoroughly settled, filtered, washed with dilute ammonium sulphide water and purified by dissolving in a small quantity of hydrochloric acid and reprecipitating with ammonium sulphide. It is then filtered off, washed as before, dried, placed in a weighed porcelain crucible, covered with a little sulphur and ignited in a current of hydrogen introduced into the crucible by a tube passing through a hole in the crucible cover. The pure manganese sulphide thus obtained is allowed to cool and weighed. The weight multiplied by .63218 gives *manganese*.

Sulphates.—500 c.c. of the clear water are slightly acidulated with hydrochloric acid, heated to boiling, and barium chloride solution added in moderate excess. The precipitate is allowed to subside completely, collected upon a filter, washed thoroughly, dried and incinerated. It is $BaSO_4$; the weight multiplied by 0.4206 gives SO_4. If the proportion of SO_4 is very low, it will be advisable to concentrate the water to one-fifth or one-tenth its bulk before precipitating.

Control. Potassium, Sodium and Lithium.—From 250 to 1000 c.c. of the water, according to the amount of solid matters present, are evaporated to dryness in a platinum dish, and the residue heated in an air bath to about 360° F., until the weight becomes constant. This determines the total solid matter in solution.

The residue is treated with a small amount of water and sufficient dilute sulphuric acid to decompose the salts present. The dish should then be covered and placed upon the water bath for five or ten minutes, after which any liquid spurted on the cover is washed into the dish, the whole evaporated to dryness and heated to redness. A few drops of ammonium carbonate solution should then be mixed with the residue, and the ignition repeated to insure the removal of the last portions of free acid. In the majority of cases the only basic elements present in considerable quantity are calcium, magnesium and sodium. The *sodium* may be determined indirectly, therefore, by calculating from the amount of Ca and Mg found, the calcium and magnesium sulphate in the residue, and subtracting their sum, together with the silica, from the total residue.

If potassium or potassium and lithium, also, are to be estimated, the filtrate from the precipitation of the SO_4 may be used, provided that sufficient of the water has been used for the estimation. For the determination of potassium and sodium in ordinary well and river waters, not less than one liter should be employed. When lithium is to be determined, it is generally necessary to use at least two or three liters. In any case, as the alkalies are to be weighed as chloride, it is advisable to precipitate the sulphates by addition of barium chloride. Unless the sulphates are to be estimated in this portion, it is unnecessary to remove the precipitate of barium sulphate so formed.

The water is evaporated to about 200 c. c., thin, pure milk of lime added in slight excess—generally from two to three c. c. will be sufficient—to the hot liquid, and the heat continued for several minutes. It is then washed into a 250

c. c. flask, disregarding the insoluble portion adhering to the dish, which, however, should be thoroughly washed, and the washings added to the flask. After cooling, the flask is filled up to the mark with distilled water, thoroughly mixed, the precipitate allowed to settle, and the liquid filtered through a dry filter. 200 c. c. of the filtrate is measured into another 250 c. c. flask, ammonium carbonate and ammonium oxalate added, filled with water up to the mark, mixed, allowed to settle, filtered through a dry filter, 200 c. c. of the filtrate measured off and evaporated to *thorough dryness* in a platinum crucible, heating very cautiously at the last stages to avoid loss by spurting. The low-temperature burner is admirably suited for this purpose. The crucible is now covered and cautiously heated to dull redness, cooled and weighed. The residue consists of potassium, lithium and sodium chlorides. It contains sometimes, also, traces of magnesium, which may be removed by treating again with lime and with ammonium carbolate and oxalate. It is frequently of advantage, in evaporating these saline solutions, to add, when the solution becomes concentrated, several c. c. of strong hydrochloric acid. This precipitates the greater portion of the salts in a finely granular condition, and renders loss by spirting less liable to occur.

If potassium and sodium chlorides only are present, they can be rapidly estimated by dissolving the weighed residue in water, determining the total chlorine by titration with silver nitrate and potassium chromate, and applying the following rule: "Multiply the quantity of chlorine in the mixture by 2.1035, deduct from the product the sum of the chlorides, and multiply the remainder by 3.6358; the pro

GENERAL QUANTITATIVE ANALYSIS. 55

duct expresses the quantity of sodium chloride contained in the mixed chloride."

The results by this method are not accurate if either the potassium or the sodium is present in relatively small amount. In such cases the following procedure may be resorted to. The weighed chlorides are dissolved in a small quantity of water, an excess of a concentrated neutral solution of platinum chloride added, evaporated nearly to dryness at a low heat on the water bath, some 80 per cent. alcohol added, allowed to stand, the clear liquor decanted off on a small filter and the residue washed in this way several times by fresh small portions of 80 per cent. alcohol. The precipitate is then washed on to the filter with alcohol, washed again with 80 per cent. alcohol, thoroughly dried and transferred as far as possible to a watch glass. The small portion on the filter is dissolved off and the solution placed in a weighed platinum dish and evaporated to dryness. The main portion on the watch glass is then added, and the whole dried to a constant weight at about 260° F., cooled and weighed. The weight thus found multiplied by .30557 gives the weight of *potassium chloride*. This subtracted from the combined weight of the chlorides gives the weight of *sodium chloride*.

Lithium, if present, is best separated before the treatment with platinum chloride. The following method, devised by Gooch, gives good results: To the concentrated solution of the weighed chlorides amyl alcohol is added and heat applied, gently, at first, to avoid bumping, until the water disappears from the solution and the point of ebullition rises and becomes constant at a temperature which is approximately that at which the alcohol boils (270° F.), the potassium and sodium chlorides are deposited and the lithium chloride is dehydrated and taken into solution. The liquid is then cooled and a drop or two of strong hydrochloric acid added to reconvert traces of lithium hydroxide in the deposit, and the boiling continued until the alcohol is again free from water. If the amount of lithium chloride is small, it will be found in the solution and the potassium chloride and sodium chloride in the residue, excepting traces which can be allowed for. If the lithium chloride exceeds ten or twenty milligrams the liquid may be decanted, the residue washed with amyl alcohol, dissolved in a few drops of water and treated as before. For washing, amyl alcohol previously dehydrated by boiling is to be used, and the filtrates are to be measured apart from the washings. In filtering, the Gooch filter with asbestos felt may be used with advantage, applying gentle pressure by the aid of the filter pump. The crucible and residue are ready for weighing after gentle heating over the low-temperature burner. The weight of the insoluble chlorides is to be corrected by adding .00041 for every ten c.c. of amyl alcohol in the filtrate, exclusive of the washings, if only sodium chloride is present; .00051 for every ten c.c. if only potassium chloride, and .00092 in the presence of both these chlorides.

The filtrate and washings are evaporated to dryness in a ~~potassium~~ crucible heated with sulphuric acid, the excess driven off, and the residue ignited to fusion, cooled and weighed. From the weight is to be subtracted, for each ten c.c. of filtrate, .0005, .00059, or .00109, according as only sodium chloride, potassium chloride, or both were present in the original mixture.

Hydrogen Sulphide.—The following method is taken from Sutton's "Volumetric Analysis:"—

GENERAL QUANTITATIVE ANALYSIS. 57

SOLUTIONS REQUIRED :—

Centinormal Iodine.—Dry, commercial iodine is intimately mixed with one-fourth its weight of pure potassium iodide and gently heated between two clock glasses by resting the lower on a hot plate. The iodine sublimes in a perfectly pure condition. It is allowed to cool under the desiccator, 1.265 grams weighed out, together with 1.8 grams of pure potassium iodide, dissolved in about 50 c. c. of water and the solution made up exactly to a liter. The liquid must not be heated, and care should be taken that no iodine vapor is lost. 1 c.c. is equivalent to .00017 H_2S. The solution is best preserved in stoppered bottles, which should be completely filled and kept in the dark. It will not even then keep very long, and should be standardized by titration with a weighed amount of pure sodium thiosulphate, which should be powdered previous to weighing, and pressed between filter paper to absorb any moisture. 50 c.c. of the iodine solution, when of full strength, will require 0.124 gram sodium thiosulphate.

Starch Indicator.—See page 35.

ANALYTICAL PROCESS :—

10 c.c., or any other necessary volume of the iodine solution, is measured into a 500 c.c. flask, and the water to be examined added until the color disappears. 5 c.c. of starch liquor are then added and the iodine solution run in until the blue color appears; the flask is then filled to the mark with distilled water. The respective volumes of iodine and starch solution, together with the added water, deducted from the 500 c. c. will show the volume of water actually titrated by iodine. A correction should be made as follows for the excess of iodine required to produce the blue color: 5 c.c. starch solution is made up with dis-

F

tilled water to 500 c. c., iodine run in until the color matches that in the test, and the volume of iodine solution so used subtracted from the figure obtained in the first titration.

Hardness. CO_3 in Normal Carbonates.—Waters containing considerable quantities of calcium and magnesium salts are said to be hard. Since the solution of calcium and magnesium carbonate in water depends partly upon the presence of carbon dioxide, boiling precipitates the greater portion of the carbonates, the result being to diminish the hardness, *i. e.*, soften the water. Magnesium and calcium sulphates and chlorides are not precipitated in this way. Hardness, therefore, is divided into two classes, temporary and permanent; the former being that which may be removed by boiling. The process herewith described is due to Hehner.

SOLUTIONS REQUIRED :—

Standard Sodium Carbonate.—1.06 grams of recently ignited pure sodium carbonate are dissolved in water and the solution diluted to 1000 c. c. 1 c.c. = .00106 gram Na_2CO_3, equivalent to .001 gram $CaCO_3$.

Standard Sulphuric Acid.—1 c. c. of pure concentrated sulphuric acid is added to about 1000 c. c. of water. 50 c. c. of the standard sodium carbonate is placed in a porcelain dish, heated to boiling, a few drops of a solution of phenacetolin or lacmoid added, and the sulphuric acid cautiously run in from a burette until the proper change of color occurs. From the figure thus obtained, the extent to which the acid should be diluted in order to make 1 c. c. of the sodium carbonate equivalent to 1 c. c. of the acid may be calculated. The proper amount of water is then added and the solution verified by again titrating.

GENERAL QUANTITATIVE ANALYSIS.

ANALYTICAL PROCESS :—

Temporary Hardness.—100 c.c. of the water tinted with the indicator is heated to boiling, and the sulphuric acid cautiously run in until the color change occurs. Each c.c. required will represent one part of calcium carbonate or its equivalent per 100,000 parts of the water; this multiplied by ten will give milligrams per liter.

Permanent Hardness.—To 100 c.c. of the water is added an amount of the sodium carbonate solution more than sufficient to decompose the calcium and magnesium sulphates, chlorides and nitrates present; usually a bulk equal to the water taken will be more than sufficient. The mixture is evaporated to dryness in a nickel or platinum dish, and the residue extracted with distilled water. The solution is filtered through a very small filter, and the filtrate and washings titrated hot with sulphuric acid as above; or 25 c.c. of distilled water may be poured on the residue, and the solution obtained filtered through a dry filter, the filtrate measured and titrated. The difference between the number of c.c. of sodium carbonate used and the acid required for the residue will give the permanent hardness.

If the water contains sodium or potassium carbonate there will be no permanent hardness, and there will be more acid required for the filtrate than the equivalent of the sodium carbonate added. From this excess the quantity of sodium carbonate in the water may be determined.

Since any alkali carbonate in the water would be erroneously calculated as temporary hardness by the direct titration, the equivalent, in terms of calcium carbonate, of the alkali carbonate present should be deducted from the figure given by the titration in order to get the true temporary hardness.

The total CO_3 in normal carbonates is given by the direct titration of the water with dilute sulphuric acid. One c.c. of the acid is equivalent to .0006 gram of CO_3.

Free Carbonic Acid.—The following process, due to Pettenkofer, is taken from Sutton's "Volumetric Analysis":—

100 c.c. of the water are put into a flask with 3 c.c. of a saturated solution of calcium chloride and 2 c.c. of a saturated solution of ammonium chloride; 45 c.c. of clear calcium hydroxide solution of known strength are added, the flask well corked, the liquids mixed, and set aside for at least twelve hours, to allow the calcium carbonate formed to settle and become crystalline and insoluble. An aliquot part (50 to 100 c.c.) of the clear liquid is then drawn off and titrated with decinormal acid, using phenacetolin or lacmoid as indicator, and from the amount required the entire proportion of calcium hydroxide unacted upon can be determined. This being deducted from the amount originally added, and the remainder multiplied by .0022, will give the weight of carbonic acid in the water in excess of that existing as normal carbonates.

Boric Acid.—*Detection.*—Add to one liter of the water sufficient sodium carbonate to render it distinctly alkaline. Evaporate to dryness, acidify with hydrochloric acid, moisten a slip of turmeric paper with the liquid and dry it at a moderate heat. In the presence of boric acid the paper will assume a distinct brown-red tint.

Estimation.—The following method for determining boric acid is due to Gooch, and has been found very satisfactory. One liter or more of the water is rendered alkaline, if necessary, by sodium carbonate and evaporated to dryness. The residue is transferred as completely as possible by the aid of slight excess of acetic acid to a flask attached to a condensing apparatus, arranged as in the

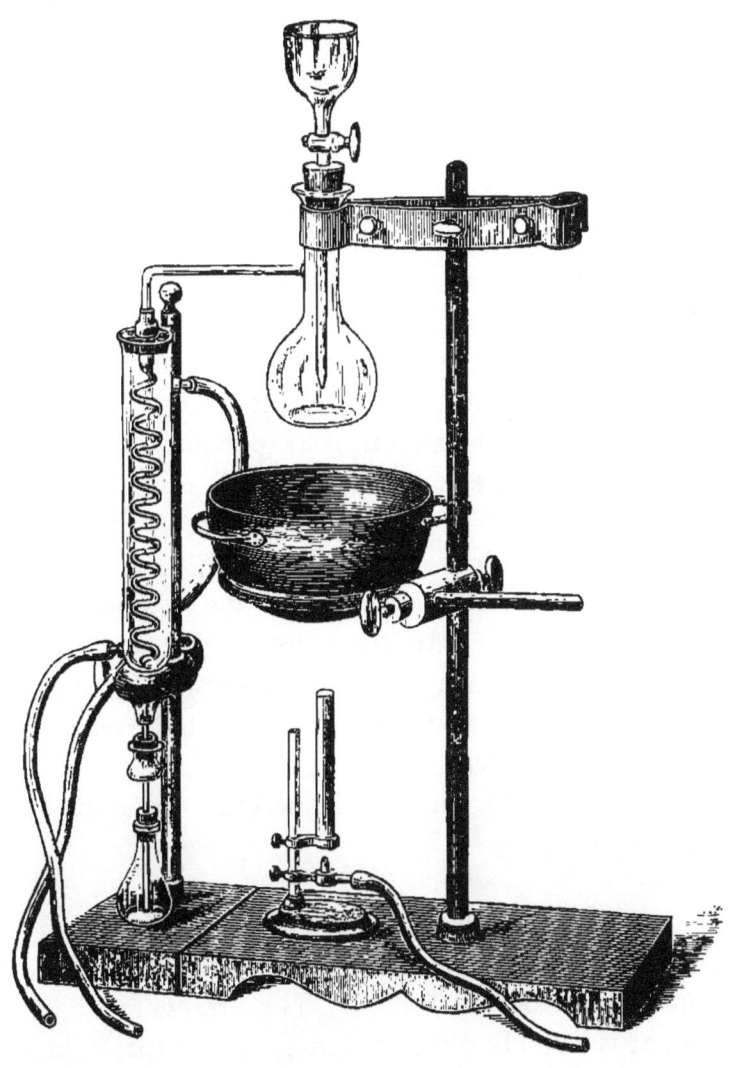

accompanying cut. About 1 gram of pure quicklime is heated in a platinum crucible over the blast lamp for from five to fifteen minutes, to decompose any hydroxide or carbonate, allowed to cool in the desiccator, and weighed. It is then introduced into the Erlenmeyer flask, slaked by the addition of a few c. c. of water, and the flask attached to the lower end of the condenser, as shown. The terminal tube of the condensing apparatus should dip into the milk of lime. The hemispherical copper basin contains paraffin, which is heated to a temperature of about 250° F. The bath is then raised so as to immerse the entire bulb of the flask, and the liquid distilled to dryness. The bath is then lowered, and when the flask and its contents have somewhat cooled, 10 c. c of methyl alcohol is introduced by means of the stoppered funnel tube, the bath raised again, and the liquid again distilled to dryness. This manipulation is repeated until six portions of 10 c. c. of methyl alcohol have been distilled off. The boric acid will distill over and be fixed by the lime. The contents of the Erlenmeyer flask are concentrated, transferred completely to the same crucible in which the lime was heated; any portions adhering to the sides of the flask or to the tube may be dissolved by a little acetic acid. The material in the crucible is cautiously dried and heated over the blast lamp for ten minutes, allowed to cool in the desiccator and weighed. The increase in weight represents boric anhydride, B_2O_3.

SPECTROSCOPIC EXAMINATION.

For the ordinary spectroscopic examination of a water a simple apparatus will suffice. The arrangement figured in the cut is a small direct-vision spectroscope, held in a universal stand, with Turquem's adjustable burner as the source of heat. The entire apparatus does not cost over $15.00, and will be found convenient and efficient.

For the examination, a liter or more should be evaporated nearly to dryness, a little hydrochloric acid being added near the end of the process, the residue placed in a narrow strip of platinum foil having the sides bent so as to retain the liquid, and heated in the flame. While this method will be sufficient in many cases, a far better plan is to separate the substance sought for in a state of approximate purity and then examine with the spectroscope. Very small traces of lithium, for instance, may be detected as follows: To about a liter of the water sufficient sodium carbonate is added to precipitate all the calcium and magnesium, and the liquid boiled down to about one-tenth its bulk; it is then filtered, the filtrate rendered slightly acid with hydrochloric acid and evaporated to dryness. The residue is boiled with a little alcohol, which will dissolve out the lithium chloride. The alcoholic solution is evaporated to dryness, the residue taken up with a little water and tested in the flame.

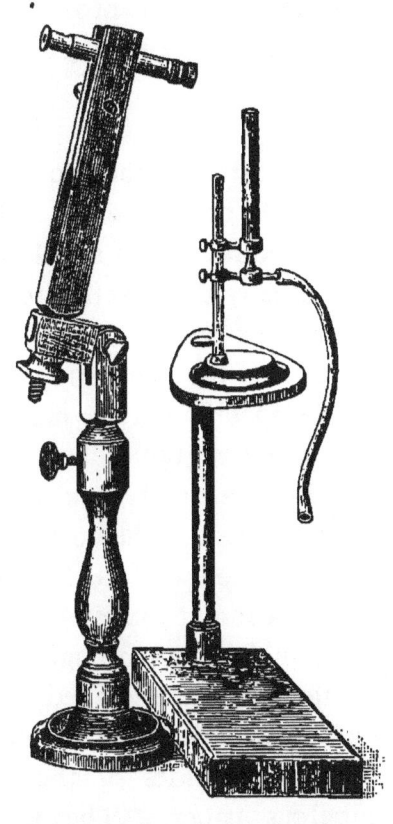

In order to identify with certainty any line which may be obtained, it is only necessary to hold in the flame at the same time a wire which has been dipped in a solution of the substance supposed to be present, and to note whether the lines produced by it and the material under examination are identical.

SPECIFIC GRAVITY.

In the great majority of cases the determination of specific gravity is not essential. Ordinary river, spring and well waters contain such small proportions of solid matter that it is usually the practice to take a measured volume and to assume its weight to be that of an equal bulk of pure water. If the proportion of solids is high, a determination of the specific gravity may be desirable. For this purpose the specific gravity bottle may be used. This consists merely of a small flask provided with a finely perforated glass stopper. The bottle is weighed first alone, then filled with distilled water at 60°, and finally with the water under examination at the same temperature. In filling the bottle, the liquid is first brought to the proper temperature, the bottle completely filled, the stopper inserted, and the excess of water, forced out through the perforation and around the sides of the stopper, carefully removed by bibulous paper. The weight of the water examined divided by the weight of the equal bulk of distilled water at the same temperature gives the specific gravity.

Another method, and one which gives very satisfactory results, is by the use of a plummet. This may conveniently consist of a piece of thick glass rod of about 10 c. c. in bulk, or of a test-tube weighted with mercury and the open end sealed in the flame. The plummet is suspended to the hook of the balance by means of a fine platinum

SPECIFIC GRAVITY.

wire and its weight ascertained. It is then immersed in distilled water at 60° and the loss in weight noted. The figure so obtained is the weight of a bulk of water equal to that of the plummet. This having been determined, the specific gravity of any water may be found by immersing in it the plummet and noting the loss in weight. This, divided by the loss suffered in pure water, gives the specific gravity.

INTERPRETATION OF RESULTS.

STATEMENT OF ANALYSIS.

The composition of water is generally expressed in terms of a unit of weight in a definite volume of liquid, but much difference exists as to the standard selected. The decimal system is very largely employed, the proportions being expressed in milligrams per liter, nominally parts per million; or in centigrams per liter, nominally parts per hundred thousand. Not infrequently the figures are given in grains per imperial gallon of 70,000 grains, or the U. S. gallon of 58,328 grains. Much more rarely grains per quart, parts per thousand, per cents or other inconvenient ratios are employed. In this work the composition is always expressed in parts per million. This ratio is practically equivalent to milligrams per liter, except in cases of waters very rich in solids, since in such waters a liter weighs notably more than one million milligrams. Factors for converting the different ratios are given at the end of the book.

From the analysis of a water it is rarely possible to ascertain the exact arrangement of the elements determined, but it is the custom to assume arrangements based upon the rule of associating in combination elements having the highest affinities, modifying this system by any inferences derived from the character or reactions of the water itself.

It has been demonstrated that, even in the case of mixtures of salts producing no insoluble substances, partial interchange of the basylous and acidulous radicles takes place. In a solution of sodium chloride and potassium sulphate, both sodium sulphate and potassium chloride will be found, as well as both of the original salts. And when the conditions are rendered more complex by the addition of other substances, it is obviously impossible to determine the exact arrangements. In view of these facts, it is preferable to express the composition of a water by the proportion of each element or radicle present. In this way a water, for instance, that might be supposed to contain K_2SO_4, will be expressed in terms of K and SO_4, respectively. In the case of bodies like CO_2 and SiO_2, which may possibly exist free in the water, their proportion is directly expressed as such. It frequently occurs, however, that the characteristics of some of the compounds in a water are sufficiently marked to indicate their presence, and there can be no objection to suggesting in connection with the analytical statement the inferences which may thus be drawn.

The organic matter, or its derived products, are best stated in terms of the nitrogen which they contain, thus permitting a comparison between the different stages of decomposition. It is inadvisable to represent the amount of unchanged organic matter in terms of oxalic acid, as has been suggested, or to express the nitrogen in terms of albumin, or any other supposititious compound.

SANITARY APPLICATIONS.

Judgment upon the analytical results from a given sample of water depends upon the class to which it belongs, and to the particular influences to which it has been subjected.

A proportion of total solids which would be suspicious in a rain or river water, would be without significance in that from an artesian well. On the other hand, a subsoil water of unobjectionable character would contain a proportion of nitrates which would be inadmissible in the case of a river or deep water. Location has also much bearing in the case; subsoil waters near the sea will be found to contain, without invoking suspicion, proportions of chlorine which would be ample to condemn the same sample if derived from a point far inland. Hence the importance of recording at the time of collection all ascertainable information as to the surroundings and probable source of the water.

Color, Odor and Taste.—Water of the highest purity will be clear, colorless, odorless, and nearly tasteless. While in some cases a decided departure from this standard may give rise to suspicion, analytical observations are necessary to decide the point. Water highly charged with mineral matters will possess decided taste; vegetable matters may communicate distinct color, but on the other hand, it may be highly contaminated with dangerous substances and yet give no indications to the senses.

Total Solids.—Excessive proportions of mineral solids, especially of marked physiological action, are known to render water non-potable, but no absolute maximum or minimum can be assigned as the limit of safety. Distilled water and waters very highly charged with mineral matter have been used for long periods without ill effects. The popular notion that the so-called hard waters conduce to the formation of urinary calculi is not borne out by surgical experience or statistical inquiry. Most urinary calculi are composed of uric acid, and are the results of disorders of the general nutritive functions. Dr. D. Hayes Agnew, in his work on Surgery, states that, so far as his experience goes, the great majority of cases of urinary calculi among the residents of Philadelphia are from the Kensington district. The water supplied to this district is softer than that supplied to the other portions of the city.

Sanitary authorities have fixed an arbitrary limit of total solids of about six hundred parts per million, but many artesian waters in constant use exceed this. An instance is found in the well on Black's Island, near Philadelphia, given in the table of analyses, which contains nearly twelve hundred parts per million, is very agreeable in taste, and has been in constant use for some years by a number of persons without any injury. The assertion that water to be wholesome must contain an appreciable proportion of total solids is also not founded on clinical experience or physiological knowledge. A discussion of the effects of special mineral ingredients, *e. g.*, magnesium sulphate, ferrous carbonate, etc., belongs to general therapeutics.

The odor produced on heating the water residue is often of much use in detecting contamination. Odors similar to those produced by heating glue, hair, rancid fats, urine, or other animal products, will give rise to grave suspicion. On the other hand, a more favorable judgment may be given when the odor recalls those given off in the heating of non-nitrogenized vegetable materials, such as wood fibre.

Poisonous Metals.—The proportion of iron in water constantly used for drinking purposes should not much exceed three parts per million. Lead, copper, arsenic and zinc must be considered dangerous in any amount, though it appears that zinc and copper, being less cumulative, are rather less objectionable in minute amount than the other two. Concerning the limit of safety with manganese and chromium but little is known, but their presence in appreciable quantity must be looked upon with suspicion.

Chlorides and Phosphates.—Chlorides—principally sodium chloride—and phosphates are abundantly distributed in rocks and soils, and find their way into natural waters; but while the former are freely soluble, and remain in undiminished amount under all conditions to which the water is subjected, all but minute amounts of the latter are either precipitated or removed by the action of living organisms. Surface and subsoil waters ordinarily contain but a few parts per million. Both chlorides and phosphates being constant and characteristic ingredients of animal excretions, it is obvious that an excess of them in natural waters, unless otherwise accounted for, will suggest direct contamination. Proximity to localities in which sodium chloride is abundant, such as the sea or rock-salt deposits, will deprive the figure for the chlorine of diagnostic value, nor can any indication of sewage or other dangerous pollution be inferred from high proportion of chlorine in deep well waters. Further, it has been shown that the proportion of chlorine in uncontaminated waters is tolerably constant, while in water subjected to the infiltration of sewage, the chlorine undergoes marked variation in amount. In most cases, therefore, a correct judgment can only be attained by comparison with the average character of the waters of the same type in the district, and by examination at intervals of the water in question.

As regards phosphates, Hehner, who has published a series of analyses, states that the presence of more than 0.6 parts per million—calculated as PO_4—should be regarded with suspicion. On the other hand, the absence of phosphates affords no positive proof of the freedom from pollution. The application of Heisch's test will often afford additional information on this point. F. E. Lott, who has applied this test in the examination of a number of waters, fully confirms Dr. Frankland's statements in regard to it, and draws the following conclusions:—

Any water which undergoes butyric fermentation when simply treated with cane sugar and kept at a temperature of 80° F., may be at once condemned as unfit for domestic use.

The single fact of a water not undergoing butyric fermentation is no proof of its purity.

A water which remains clear under this treatment would certainly be less likely to be contaminated by sewage than one which becomes milky, and the possibility of unoxidized sewage matter being in a water which remains quite clear is very doubtful.

The butyric ferment is not perceptibly influenced by the presence of abnormal amounts of chlorine, ammonia, nitrogenous matter, sulphates or nitrates, but is a very accurate indicator of the presence of phosphates.

Nitrogen from Ammonium Compounds.—Ammonium compounds are usually the result of the putrefactive fermentation of nitrogenous organic matter; they may also be the product of the reduction of nitrites and nitrates in presence of excess of organic matter. In either case, therefore, they suggest contamination. Deep well waters nearly always contain an excess of ammonium compounds, derived, in large part, from the decomposition of nitrates. Their presence here is hardly ground for adverse judgment, since the water, even though originally contaminated, has undergone extensive filtration and oxidation and its organic matter converted into bodies presumably harmless. Such waters, indeed, usually show only traces of unchanged organic matter.

Rain water often contains large proportions of ammonium compounds; but here, also, the fact cannot condemn the water, since it does not indicate contamination with dangerous organic matter.

The evolution of ammonia in the distillation of rain water often continues indefinitely, the larger portion passing over in the first distillates, but small quantities being present even after the distillation has been much prolonged. The same continuous evolution of ammonia is noted in waters containing urea, but in this case a larger proportion is collected in the earlier distillates, nearly all coming over before one-half the water has been distilled. Fox gives the following figures as ratios obtained in the analysis of two samples, one of rain water collected from a roof and therefore impure, and the other of a water containing urine :—

		RAIN WATER.	URINE WATER.		
1st distillate,	. . . .	.35	.38	parts per million.	
2d "	. . . .	.25	.14	"	"
3d "	. . . .	.12	.065	"	"
4th "	. . . .	.09	.035	"	"
5th "	. .	.09	. .	"	"
6th "	. . . .	.04	. .	"	"
7th "	. . .	.03	. .	"	"
		.97	.620		

Nitrogen by Alkaline Permanganate (Nitrogen of "albuminoid ammonia").—A large yield of ammonia by boiling with alkaline potassium permanganate will, of course, point to an excess of nitrogenous organic matter.

The inferences to be drawn depend upon the origin and condition of the organic material. If animal, the water may at once be condemned as unsafe. Waters containing excessive amounts even of vegetable matter are not free from objection, since they have frequently caused persistent diarrhœa. If the organic matter, whether animal or vegetable, is in a state of active decomposition, it is doubly objectionable. Mallet has called attention to the fact that such waters, as a rule, yield ammonia rapidly, whereas non-decomposing material yields it but slowly, and he points out the importance, therefore, of noting the rate at which the ammonia collects in the distillate.

Dr. Charles Smart states that water containing fermenting vegetable matter is colored yellow by boiling with sodium carbonate, and that when Nessler's reagent is added to the distillate a greenish, in place of the ordinary yellowish-brown, color is produced. He applies this fact in conjunction with the determination of the oxygen-consuming power (Tidy's process) and the rate of evolution of the ammonia by alkaline permanganate as follows:—

A water yielding ammonia slowly by alkaline permanganate, contains recent organic matter; of animal derivation, if the oxygen-consuming power is low; of vegetable, if high.

A water yielding ammonia more rapidly by alkaline permanganate shows decomposing organic matter; of animal origin, if the oxygen-consuming power is low and there is no interference with the Nessler reaction; of vegetable origin, if the oxygen consumed is high, and if a yellow color is produced in the water by sodium carbonate, and a greenish color in the Nesslerized distillate.

Inferences as to the source of the organic matter can

be more safely drawn from the amount of chlorine and nitrates present. If the chlorine is high, *i. e.*, in excess of the average of the district, it may reasonably be inferred that the material is, at least in great part, of animal origin. In this case the nitrates will usually either be high or entirely absent, according as the contaminating matter has passed through soil or enters the water directly.

It is well to note that a large amount of vegetable matter will, as a rule, show itself by the color it imparts to the water.

Wanklyn gives the following standards:—

High purity, ...00 to .041 per million.
Satisfactory purity, ...041 to .082 " "
Impure, ... over .082.

In the absence of ammonium compounds, he does not condemn a water unless the nitrogen by permanganate exceeds .082 per million; but a water yielding 0.123 parts per million of nitrogen by permanganate he condemns under all circumstances.

Nitrogen as Nitrites.—Nitrites are present in water as the result either of incomplete nitrification of ammonia, or the reduction of already formed nitrites, under the influence of excess of organic matter. Since they are transition products less stable than either of the extremes, their presence in water is usually evidence of existing fermentative changes, and, further, may be taken as indicating that the water is unable to dispose of the organic contamination. When, however, the conditions are such that oxidation cannot take place, nitrites may persist for a long time. This occurs in deep waters in which fermentative changes have long since ceased, but oxygen is not available. These contain not infrequently small amounts of nitrites, to which the same degree of suspicion cannot be attached. When nitrites are found in these waters, the possibility of their introduction from polluted subsoil water through defective tubing must not be overlooked. Rain water, also, sometimes contains nitrites derived from the air, and therefore not indicative of any putrefactive change. The presence of measurable quantities of nitrites in river or subsoil water is sufficient grounds for condemnation.

Nitrogen as Nitrates.—Nitrates, the final point in the oxidation of nitrogenous organic matter, are normally present in all forms of water, but only in limited amounts. They are derived, in by far the larger part, from the oxidation of animal matters. Rain water and that from mountain streams and deep wells, except from cretaceous strata generally contain only traces, but river and subsoil water will always contain appreciable amounts, unless some destructive action, such as recent sewage pollution, is at work. When, therefore, a water contains enough mineral matter to demonstrate its percolation through soil, and at the same time is free from nitrates or contains only traces, the occurrence of a destructive fermentation may be inferred. These cases are not uncommon among well waters, and the samples are generally not clear. Decided departure, either by increase or decrease, from the proportion of nitrates usual in the waters of any district must be taken as evidence of contamination.

Oxygen-consuming Power.—Sanitary authorities differ very much as to the significance of this datum. Attempts have been made to fix maximum limits for the various types of water, and also to gauge the character and condition of the organic matter by observing the rate at which the oxidation takes place, but no positive conclusions can be given. In general, it may be said that a sample which has high oxygen-consuming power will be more likely to be unwholesome than one which is low in this respect; but the interferences are so numerous, and the susceptibility to oxidation of different organic matters of even the same type, is so different, that the method is at best only of accessory value.

The following proportions are given by Frankland and Tidy as the basis for interpreting the results of this method:—

OXYGEN ABSORBED IN THREE HOURS.

High organic purity, ... 0.5   parts per million.
Medium purity, ... 0.5 to 1.5   "   " . "
Doubtful, ... 1.5 to 2.1   "   "   "
Impure, ... over 2.1   "   "   "

Dissolved Oxygen.—Full aëration of water is favorable to the destruction of organic matter, and a decided diminution of the normal quantity of dissolved oxygen may show excess of such matter and of microbic life. Gérardin has pointed out that this diminution is associated with the development of certain low forms of vegetable life, and Leeds has recorded similar facts. These changes are more likely to take place in still waters, and are frequently accompanied by disagreeable odor and taste. In cases in which stored waters become unpalatable these facts should be borne in mind.

Dupré has given the following as the basis for interpreting the results of his adaptation of the determination of dissolved oxygen:—

"A water which does not diminish in its degree of aëration during a given period of time, may or may not contain organic matter, but presumably does not contain growing organisms. Such organic matter as it may be found to con-

RESULTS FROM UNCONTAMINATED WATERS. 77

tain by chemical analysis need not be considered as dangerous impurity."

"A water which by itself, or after the addition of gelatin or other appropriate cultivating matter, consumes oxygen from the dissolved air, at lower temperatures, but does not consume any after heating for say three hours at 140° F., may be regarded as having contained living organisms, but none of a kind able to survive exposure to that temperature."

"A water which by itself, or after addition of gelatin or the like, continues to absorb oxygen from the contained air after heating to 140° F., may be taken as containing spores or germs able to survive that temperature."

Hardness.—The degree of hardness has but little bearing on the sanitary value of water, but is important in reference to its use for general household purposes, in view of the soap-destroying power which hard waters possess.

USUAL ANALYTICAL RESULTS FROM
UNCONTAMINATED WATERS.

Parts per million.

	Rain.	Surface.	Subsoil.	Deep.
Total solids, ...	5 to 20	15 upward	30 upward	45 upward
Chlorine, ...	Traces to 1	1 to 10	2 to 12	Traces to large quantity
Nitrogen by permanganate,	.08 to .20	.05 to .15	.05 to .10	.03 to .10
" as NH$_4$, ...	.20 to .50	.00 to .03	.00 to .03	Generally high
" " nitrites, ...	None or traces	None	None	None or traces
" " nitrates, ...	Traces.	.75 to 1.25	1.5 to 5	.00 to 3

ACTION OF WATER ON LEAD.

The almost universal use of lead pipes for conveying water and the facility with which some waters corrode and dissolve the metal, makes it a question of moment to determine the cause of this action and to devise means for its prevention. The subject has received considerable attention within the last few years, and the conditions which determine corrosion are now fairly understood. As a rule, it is found that waters free from mineral matter dissolve lead with facility, and especially in the presence of oxygen. Some very soft waters, however, are entirely without action, and the fact remained unexplained until a few years ago, when Messrs. Crookes, Odling and Tidy found that the action was controlled by the amount of silica contained in the water. They found that: Those soft waters which, when taken from the service pipes, contained a notable quantity of lead, gave, on the average, three parts of silica per million; in those in which there was no lead, the silica present amounted to 7.5 per million and in those in which the action was intermediate, 5.5 parts per million. That it was really the silica that conditioned the corrosion was confirmed by laboratory experiments. They also found that the most effective way of silicating a water is by passing it over a mixture of flint and limestone. The reason for this was pointed out later by Messrs. Carnelly and Frew, who showed that while calcium carbonate and silica both exert a protective influence, calcium silicate is more effective than either, and further, that in almost all cases in which corrosion took place it was greater in the presence of oxygen. This is particularly the case with potassium and ammonium nitrates and with calcium hydroxide. The reverse is true of calcium sulphate, which is more corrosive when air is excluded. Their experiments also show that the presence of calcium carbonate and calcium silicate altogether prevent corrosion by potassium and ammonium nitrates.

As the result of an elaborate series of experiments, Müller concludes, that while chlorides, nitrates and sulphates all act upon lead pipes, no corrosion takes place in the presence of sodium acid carbonate, and that calcium carbonate, by taking up carbonic acid, acts in the same way. This latter conclusion is opposed by the experiments of Carnelly and Frew, who found calcium carbonate equally effective when carbonic acid is excluded. Müller also states that surface waters contaminated by sewage and containing large amounts of ammoniacal compounds will dissolve lead under all circumstances.

Allen has shown that water containing free acid, including sulphuric acid, acts energetically upon lead. This is not surprising in view of the later experiments, which prove that even calcium sulphate is corrosive. Later, W. Carleton-Williams found that even in the presence of free acid corrosion may be prevented by the addition of sufficient silica. His experiments also confirm the view generally held, that soluble phosphates protect lead to a marked degree.

The following is a summary of the more important facts in regard to this suubject :—

Corrosive : Oxygen, free acids or alkalies, nitrates, particularly potassium and ammonium nitrates, chlorides and sulphates.

Non-corrosive and preventing corrosion by the above: Calcium silicate, calcium carbonate, sodium acid carbonate, ammonium carbonate, silica and soluble phosphates.

LIVING ORGANISMS IN WATER.

In a comprehensive sense the living organisms of water include representatives of all the great groups of animals and plants. The presence of any of the higher orders of organic forms may be taken generally as an indication of, at least, moderate purity, as these are absent from very foul water. From an analytical point of view, observation is limited to the determinations of those forms which are inappreciable to the unassisted eye. As far as regards some of the moderately complex organisms, such as the minute crustaceans, algæ, desmids, and even the amœbæ, it may be said that while some general inferences as to the character and source of the water may be deduced from an identification of the specific forms, no definite sanitary signification can be attached to them. The ova of the entozoa might in some cases be detected by careful search, and these would, of course, be of much importance as indicating recent pollution of a highly dangerous character. Modern investigation is practically limited to the determination of the minute forms of life, essentially vegetable, included under the terms micrococci, bacilli, spirilli, bacteria, etc., known collectively as micro-organisms or microbes. These are, in some cases, connected with the transmission of infectious diseases. The recognition of the individual species of microbes by microscopic examination of water-samples would be a very difficult task. The method now generally used is due mainly to Koch, and is known as plate-culture. It depends on the fact that the specific differences of many microbes are distinctly indicated by their method of aggregation in multiplication. A small, accurately measured volume of water is mixed with a large volume of nutritive (culture) fluid, spread over a glass plate, and placed under conditions favorable to the growth of micro-organisms. The principle of the process is, that by the dilution and extension of surface each living microbe becomes a centre of development of a colony of its own kind, and thus a close approximation of the number present in the original sample, together with the nature of the prevailing specific forms, may be obtained. This method has been the subject of many experiments within the past few years, but it cannot be said to be yet so far perfected as to furnish an absolute test for the purity of water. Several difficulties interfere with the interpretation of results. All natural waters contain microbes, the proportion being subject to great variation, without corresponding variation in sanitary quality. The majority of forms existing in water are harmless in any proportion, and no definite method is known to distinguish these from forms of pathogenic (disease producing) character. The culture fluid is usually prepared with gelatin, and those microbes which, in their growth, cause a decomposition, or, as it is technically termed, "liquefy the gelatin," are regarded as objectionable, but this does not appear safe as an absolute rule. The proportion of living microbes in a water is subject to rapid increase for a brief period after collection of the sample, and may be greatly modified by incidental conditions during storage or transportation, so that little value can be attached to quantitative determinations, except when made without appreciable delay. The culture fluids used, and the conditions under which the cultivation takes place, do not suffice for the development of all the forms present. The cultivation ought to be extended over many days and different samples of the same water tried, with various nutritive media, and at various temperatures, to secure a full knowledge of the microbes present. In judging of the value of these methods it must not be overlooked that the entire alimentary tract is abundantly charged with micro-organisms, both as to number and to variety; and further, that while research has shown that some forms are frequently, if not invariably, associated with particu'ar diseases, a direct causative relation has in most of these cases not been established.

While, then, these so-called bacteriological examinations are as yet of uncertain value in the determination of the potability of water, they have been of much use in determining the effects and usefulness of certain conditions to which water is subjected. In these studies the method is sufficiently free from fallacy to make the results trustworthy. By it, it has been shown that filtration at first greatly diminishes the number of micro-organisms, but subsequently, owing to the fouling of the filter, and partly to the penetration of successive colonies of microbes through the pores of the filter, the filtrate becomes richer in microbes than the unfiltered water. When suspended mineral matters are caused to settle, a large proportion of the microbes is carried down; but if the water thus purified is not soon removed, the microbic life again develops, it may be even in greater proportion than was originally present. The storage of water at first increases and then diminishes the proportion present. The presence of free acid, even of carbonic acid, is decidedly inhibitory to the development. When microbes essentially foreign to the water are introduced they are often soon destroyed, apparently under the influence of those forms naturally present, and, therefore, better adapted to existing conditions; but this is by no means always the case, some of the more virulent pathogenic organisms having high resisting power.

PURIFICATION OF DRINKING WATER.

The most obvious method of purifying water is by distillation. The process is too expensive for general use, but is especially adapted for water intended for pharmaceutical or chemical purposes. It has also been used for supplying vessels at sea and in tropical localities in which the natural waters may be contaminated with malarial or other germs.

The methods in general use for purifying water are simple filtration and the removal of the impurities by appropriate chemical agents. Dr. Plagge, of the Hygienic Institute of Berlin, has examined a number of filters for household use, and obtained the following results:—

Carbon filters not only permitted the free transmission of micro-organisms, but in some cases the number of these in the filtrate greatly exceeded that in the unfiltered water. In one case, the unfiltered water yielded 68 colonies per c.c.; the filtered, 12,000. Frankland also found a similar result, and, as he remarks, the filtering medium obviously acts as a hotbed for the development of the micro-organisms.

The stone and sand filters were all found to be worthless.

The spongy iron filter yielded the following results:— The unfiltered water yielded 34,000 colonies per c.c.; the filtered water, 18,000 to 24,000.

The paper filters yielded very unsatisfactory results.

The earthenware filters on Pasteur's principle gave, in nearly every instance, a filtrate practically free from micro-organisms. Thus in one case the unfiltered water yielded 284, and the filtered only 4 colonies per c.c.

The asbestos filters gave results essentially similar to those obtained with the earthenware filters.

The advantage of the earthenware and asbestos filters is, that they will bear treatment with boiling water, or the application of dry heat, by which any organic life which may penetrate the pores or lodge closely on the surface, can be destroyed.

For the purification of drinking water on the large scale, sand filter beds have been found to be more efficient; but good results are obtained only under proper supervision and attention to cleanliness and renewal of the filtering material.

For the purification of hard waters, any of the methods detailed below may be applied, and the treatment by such means has been found of great advantage. Dr. Percy Frankland, as the result of an extended series of careful investigations, comes to the following conclusions:—

" Organized matter is, to a large and sometimes to a most remarkable extent, removable from water by agitation with suitable solids in a fine state of division, but such methods of purification are unreliable.

" Chemical precipitation is attended with a large reduction in the number of micro-organisms present in the waters in which the precipitate is made to form and allowed to subside.

" If subsidence either after agitation or after precipitation be continued too long, the organisms first carried down may again become redistributed throughout the water."

It is essential, therefore, that the liquid be filtered as short a time after the precipitation as possible.

If a small quantity of alum is added to a water, it is decomposed with the formation of a flocculent precipitate of aluminum hydroxide, which settles comparatively rapidly, and carries down with it all suspended matters as well as a large proportion of the dissolved organic matters. Waters which contain such an excess of organic matter as to be distinctly colored, may be made quite clear and colorless by this treatment. Two grains of alum to the gallon is sufficient for the purpose, but if very rapid subsidence is desired more may be added.

The addition of a ferric salt to water is attended with the same decomposition and formation of a precipitate as with alum, and the reaction has been utilized with great advantage in the purification of water. One of the methods proposed was by treating the water with spongy iron, by which a certain amount of the iron is dissolved and is subsequently removed by passing the water through sand filter beds. It was found, however, that the iron soon lost its activity, by reason of the formation of rust, which prevents further action. This difficulty is removed by the use of granular iron contained in an iron cylinder [Anderson and Ogsten, *Proc. Inst. Civ. Eng.*, Vol. 81], which is rotated while the water passes through ; the iron is brought into thorough contact with the water, and there is sufficient abrasion to keep its surface clean. The time of contact with the iron of course depends upon the relative impurity of the water. For Antwerp water, which is purified by this means, the maximum effect is accomplished in 3.5 minutes. After leaving the cylinder, the water is passed through sand filter beds in which the iron is oxidized and removed. Examination of the purified water shows it to be practically sterilized, and the quantity of nitrogen obtainable by potassium permanganate is reduced to from one-half to one-third the amount which the water originally furnished.

The experiments of Dr. P. Frankland have shown that micro-organisms, at least those which are capable of development in gelatin solutions, may be completely removed by the use of powdered coke as a filtering material. With this, as with all similar materials, the best results are only secured by frequent renewal and a not too rapid rate of filtration. Messrs. Salamon and Matthews [*J. Soc. Chem. Ind.*, 1885, p. 261] point out that the coke at the same time effects a decided reduction in the amount of organic matter in the water. They found, further, that the action of the coke is due mainly, if not entirely, to the presence of metallic iron.

IDENTIFICATION OF THE SOURCE OF WATER.

The determination of the course of underground streams, and of communications between collections of water, is often an important practical problem. In geological and sanitary surveys, valuable information may occasionally be gained. The method generally pursued when connection between water at accessible points is to be detected, is to introduce at one point some substance not naturally existing in the water, and capable of recognition in small amount. Lithium compounds are among the best for this purpose. They are not frequent ingredients of natural waters, and are easily recognized by the spectroscope. Lithium chloride is the most suitable. The quantity to be employed will vary with circumstances. It scarcely needs to be stated that the waters under examination should be carefully tested for lithium before using the method.

When the lithium method is inadmissible, recourse must be had to other substances of distinct character, such as strontium chloride, but this possesses the disadvantage that a considerable amount may be rendered insoluble, and thus lost in the ordinary transit through soil. Recently, use has been made of the newly discovered organic coloring matters of high tinctorial power, one of the most suitable of which is *fluorescein*, $C_{20}H_{12}O_5$, a derivative of benzene. This will communicate a characteristic and intense fluorescence to many thousand times its weight of water. An entire river may be colored by a few kilograms. By its use an underground communication was demonstrated to exist between the Danube and the Ach, a small river which flows into the Lake of Constance.

A more important feature of the problem in a sanitary point of view is the determination of the source of a given current or collection of water, when such source is inaccessible. Problems of this character are not infrequent in large cities in which the systems of water supply and drainage are defective, thus giving occasion to accumulations of water in cellars and similar places. Often, in these cases, no extended explorations can be made by reason of the adjacent buildings and conflicting property interests, and the question may arise whether the water proceeds from a leaky hydrant, drain, sewer, or subsoil current. It is obvious that in the case of the collection of water in a cellar from causes other than surface washings or entrance of rain, it must have passed through some distance of soil, and in built-up districts will almost certainly be charged with organic refuse. To correctly interpret the results, it will be necessary to know the general character of the subsoil water of the district and the composition of the city supply. As a rule, the transmission of water through moderate distances of soil will not materially increase the mineral constituents. Hence, if the sample contains an excess of dissolved matters as compared with the water supply of the district, it may reasonably be inferred that it is derived from a drain, sewer, or subsoil current.

In these investigations it will generally be sufficient to

determine the total solids, odor on heating, chlorine, nitrates and nitrites. The following figures are from some results obtained in investigations made in association with Mr. Chas. F. Kennedy, Chief Inspector to the Board of Health of this city:—

	CITY SUPPLY.	CELLAR WATER.		
		No. 1.	No. 2.	No. 3.
Total solids,	115	140	661	640
Odor on heating,	faint	faint	strong	urinous
Chlorine,	4	6.4	77.0	128.0
N as ni'rates,	0.7	1.0	3.5	none
" " nitrites,	none	present	present	none

Sample No. 1 was taken from a cellar in which a small amount of water had been almost constantly present for a long time, and of which the source could not be ascertained. The results of analysis led to the view that since it resembled in composition the city supply, it was derived from a leaky hydrant pipe. The parties in interest were not inclined to accept this opinion, but the examination of the condition of the hydrant on an adjacent property, showed a leak, which being repaired the water ceased to appear in the cellar. In this case it was found that the water had passed through twenty-two feet of earth. In the second case the sample is seen to be very impure, and it was suggested that it was derived directly from a leaky drain, which upon exploration proved to be the case. In the third sample, the high chlorine, strong urinous odor and absence of nitrates and nitrites pointed unmistakably to recent and profuse con'amination with foul water.

Occasionally the analytical results will be ambiguous, and it is advisable to make examinations of more than one sample, since accidental circumstances, rain-fall, etc., may affect the composition of the water.

Instances of the contamination of water by unusual substances are occasionally noted, and these sometimes afford a clue to the source of the water. Among the instances of this kind within our own experience may be noted the contamination with petroleum and with soap. In the former case it was evident that the contamination was from a leaky pipe connecting two refineries. In the latter it was shown to be derived from an adjoining building used as a laundry.

TECHNICAL APPLICATIONS.

Boiler Waters.—The main conditions affecting the value of a water for steam-making purposes are its tendency to cause corrosion and the formation of scale. *Corrosion* may be due to the water itself, to the presence of free acids, or to substances which form acids under the influence of the heat to which the water is subjected. Pure water, *e.g.*, distilled water, exhibits a powerfully corrosive action upon iron. The dissolved oxygen which all waters contain also aids in the corrosion, and especially when accompanied, as is usually the case, by carbonic acid. There is always greater rusting at the point at which the water enters the boiler, since there the gases are driven out of solution and immediately attack the metal. This is an evil that obtains with all waters, and it is not customary, in making examinations for technical purposes, to determine the amount of these bodies. In water that has had free access to air, the oxygen in solution is a tolerably constant quantity, and it is sufficient to note the temperature and refer to the table of amounts of oxygen dissolved in water. The corrosive action of oxygen and carbonic acid is especially noticeable in waters that are comparatively pure, such as those derived from mountain springs. This was repeatedly observed by one of us in the examination of the waters used for the locomotives of the Baltimore and Ohio R.R. The waters which caused the most corrosion were mainly those containing small quantities of solid matter, the full amount of oxygen and considerable carbonic acid, but no other acid or acid forming body.

Free acid, other than carbonic acid, is not often found in water, and if present, renders the water unfit for use, unless it be neutralized. Mine waters are the most likely to contain free acid, sulphuric acid being generally present. Sometimes the acidity is due to organic acids. These act very injuriously on iron. Allen gives an interesting example of this in the water supplied to Sheffield, Eng., which he found to contain an organic acid in amount equivalent to from 3.5 to 10 parts of sulphuric acid per million.

Magnesium chloride is frequently present in waters, and if in considerable quantity may be very harmful. At a temperature of 310°F., corresponding to an effective pressure of four atmospheres, magnesium chloride reacts with water to form magnesium oxide and hydrochloric acid, the latter attacking the boiler, especially at the water line. If there is present at the same time considerable calcium carbonate the evil may be somewhat lessened, but as Allen has pointed out, and we also have noticed, there may still be corrosion, so that the presence of more than a small quantity of the salt, say a grain or two to the gallon, may be considered objectionable. Allen remarks that the presence of a certain amount of sodium chloride may prevent this decomposition, the two chlorides combining to form a stable double salt. The addition, therefore, of common salt to a water containing magnesium chloride may act to diminish corrosion, a point which will bear further investigation.

It has not been determined how far the presence of nitrites, nitrates and ammonia affects the quality of water for steam-making purposes; but it is more than probable that they act harmfully, especially the nitrates, which are frequently present in large amount.

Scale is composed of matters deposited from the water either by the decompositions induced by the heat or by concentration. When the deposit is loose it is termed *sludge* or *mud*, and usually consists of calcium carbonate, magnesium oxide and a small amount of magnesium carbonate. The magnesium oxide is formed by the decomposition of the magnesium carbonate and chloride. This fact was first pointed out by Driffield (*J. Soc. Chem. Ind.*, VI, 178).

The formation of sludge is the least objectionable effect, since it may readily be removed by "blowing off," provided that care is previously taken to allow the flues to cool down so that when the water is removed the heat of the flues may not bake the deposit to a hard mass. Waters containing calcium sulphate form hard incrustations difficult to remove and causing great loss of fuel by interfering with the transmission of the heat to the water. It not only forms a hard incrustation in itself, but becomes incorporated with the mud and renders it also hard. The hard scale will also contain practically all the silica and the iron and aluminum present in the water, besides any matters originally held in suspension.

It follows from the above that a water only temporarily hard, will, if care is taken in the management of the boiler, cause the formation merely of loose deposit of sludge—temporary hardness being due in the main to calcium and magnesium carbonates. A water permanently hard will probably form a hard scale, since such hardness is usually due to calcium sulphate.

In accordance with these principles, the analysis of a water for steam-making purposes may include the determinations of free acid, other than carbonic acid, total solid residue, SO_4, Cl, Ca, Mg, temporary and permanent hardness. In cases in which the qualitative tests show but small amounts of SO_4 and Cl, the analysis may be limited to the determinations of the temporary and permanent hardness.

It has been pointed out in an earlier chapter that it is not possible to deduce from the analytical results the exact forms in which the various elements are combined, but since it is known that at the high temperature ordinarily reached in boilers definite chemical changes occur, it is safest to exhibit the maximum amount of corrosive and scale forming ingredients which the water under these circumstances could develop. Thus, since calcium sulphate is practically insoluble in water above $212°$ F., the proportion of calcium sulphate may be regarded as such as would be formed by the total quantity of calcium or the total quantity of SO_4, according to which is present in the larger amount. Similarly, as the decomposition of magnesium chloride is promoted by the high temperature of the boiler, the analytical statement should indicate the maximum proportion of this compound obtainable from the magnesium and chlorine present. These rules cannot apply absolutely to waters rich in alkali carbonates, since these would neutralize any acid formed from the magnesium chloride, or even prevent its formation, and would prevent to a large extent the formation of calcium sulphate. Much remains to be determined concerning the effects of the high temperature and concentration to which boiler waters are subjected.

Purification of Boiler Waters.—The problems presented in the treatment of boiler waters are usually the removal of the calcium carbonate and sulphate, and magnesium carbonate and chloride. Both carbonates are appreciably soluble in pure water. About one grain of calcium carbonate to the gallon is usually stated to be the proportion dissolved, but it has been pointed out lately by Allen that this is an understatement, since solutions have been obtained containing twice this amount. If the water contains carbonic acid it will take up a much greater proportion of the carbonates, but in this case they will be deposited from the solution by boiling. This has been accounted for by supposing the existence of a soluble bicarbonate, $CaCO_3H_2CO_3$, which is decomposed by the boiling. The facts do not support this supposition. It is found that the amount of carbonic acid in the water bears no relation to the amount of dissolved carbonate. Allen thinks it possible that calcium carbonate exists in two forms, one being crystalline and insoluble, and the other colloid and soluble, and that the soluble modification may be converted into the insoluble by simple boiling. Be this as it may, the greater proportion of these carbonates can be thrown out of solution by any means that will deprive the water of the carbonic acid. Sodium hydroxide is usually the best for the purpose, and should be added in quantity just sufficient to form normal sodium carbonate. If there are present in the water calcium and magnesium chlorides and sulphates, these also will be decomposed and precipitated by the sodium carbonate so formed. If the amount of sodium carbonate is not sufficient to decompose all of these bodies, a sufficient quantity should be added with the sodium hydroxide to effect the complete decomposition. The precipitate is allowed to settle or filtered off.

In cases in which the feed water is heated before it enters the boiler, it may only be necessary to add to the water sodium carbonate in quantity sufficient to decompose the calcium and magnesium chlorides and sulphates, since the heat alone will suffice to throw down the carbonates.

Care should be taken in these precipitations that no more sodium hydroxide is added than is sufficient for the decomposition, since any excess would tend to corrode the boiler.

Clark's process consists in treating the water with calcium hydroxide (lime-water). This precipitates the calcium and magnesium carbonates by depriving the water of its free carbonic acid. It has, of course, no effect upon the calcium sulphate. It is to be noted that the proportion of calcium hydroxide which is to be added must be calculated from the amount of free carbonic acid existing in the water, and not from the amount of carbonates to be removed. The precipitate will usually require at least twelve hours for complete subsidence, but after three or four hours the water will be sufficiently clear for some purposes. If a filter press is used, as in Porter's process, the time required for clarification is very much shortened. Another advantage of this process is the use of a solution of silver nitrate, in order to determine more conveniently the proportion of calcium hydroxide which is to be employed. The lime is first slaked and dissolved in water, and the water to be softened run in and thoroughly mixed with it. From time to time small portions are taken out in a cup and a few drops of a solution of silver nitrate added. As long as the lime is in excess a brownish coloration is produced. When this has become quite faint, and just about to disappear, the addition of the water is discontinued, and, after a short time, the water is filtered by means of the press.

Soluble phosphates added to a water, precipitate completely in a flocculent condition any calcium, magnesium, iron or aluminum. This reaction can be best applied by using the tri-sodium phosphate ($Na_3PO_4 + 12H_2O$), which is now a commercial article. By reason of the facility with which this substance loses a portion of its sodium to acids, it acts not only as a precipitant to the above materials, but will neutralize any free mineral acid present in the water. From evidence submitted by those who have used the process on the large scale, it appears that not only is no hard scale formed, but that scale already existing prior to its use is gradually disintegrated and removed with the sludge. Experiments indicate that no injury results from an excess of the material; but the economical employment of the method, especially with very hard waters, can only be based upon a correct analysis, and an estimation of the phosphate required for the precipitation. In many cases the composition of the water will be such that a partial precipitation will be sufficient for all purposes.

In regard to the quality of water for technical other than steam-making purposes, such as brewing, dyeing, tanning, etc., no detailed methods or standards can be laid down. The nearest approach to purity that can be secured in the supply will be of the greatest advantage. The more objectionable qualities will be large proportion of organic matter, especially if it distinctly colors the water, and excessive amounts of iron or free acid. For the removal of these the same processes are applicable as are indicated in the directions for the purification of the water for other purposes.

ANALYTICAL DATA.

CONVERSION TABLE.

Parts per 100,000	× .7	=	Grains per Imperial Gallon	
" " 1,000,000	× .07	=	" " " "	
" " 100,000	× .583	=	" " U. S. "	
" " 1,000,000	× .058	=	" " " "	
" " 1,000,000	× .00833	=	Av. pounds per 1000 U. S. Gal.	
Grain " Imp. gallon	÷ .7	=	Parts per 100,000	
" " " '	÷ .07	=	" " 1,000,000	
" " U. S. "	÷ .583	=	" " 100,000	
" " " "	÷ .058	=	" " 1,000,000	

DIBDIN'S TABLE OF OXYGEN DISSOLVED BY WATER AT VARIOUS TEMPERATURES. EXTENDED TO GIVE THE WEIGHT OF OXYGEN PER LITER. CORRECTED TO 0°C. AND 760mm. PRESSURE.

TEMPERATURE FAHRENHEIT.	TEMPERATURE CENTIGRADE.	CUBIC INCHES OF OXYGEN PER GALLON (7000 GRAINS).	MILLIGRAMS OF OXYGEN PER LITER.
41°	5.00°	2.101	10.84
42	5.55	2.074	10.72
43	6.11	2.048	10.57
44	6.66	2.022	10.45
45	7.22	1.997	10.30
46	7.77	1.973	10.18
47	8.33	1.949	10.06
48	8.89	1.927	9.94
49	9.44	1.905	9.83
50	10.00	1.884	9.72
51	10.55	1.864	9.61
52	11.11	1.844	9.51
53	11.66	1.826	9.42
54	12.22	1.808	9.33
55	12.77	1.791	9.24
56	13.33	1.775	9.15
57	13.89	1.760	9.08
58	14.44	1.746	9.01
59	15.00	1.732	8.94
60	15.55	1.719	8.87
61	16.11	1.706	8.80
62	16.66	1.695	8.74
63	17.22	1.683	8.68
64	17.77	1.674	8.64
65	18.33	1.667	8.60
66	18.89	1.660	8.56
67	19.44	1.652	8.52
68	20.00	1.644	8.48
69	20.55	1.639	8.45
70	21.11	1.634	8.43

The table is calculated for a barometric pressure of 760 mm., and would require corrections for variations from this, but such corrections are mostly within the limits of experimental error.

ANALYSES OF RAIN AND SUBSOIL WATERS.
PARTS PER MILLION.

	From.	Total Solids.	Chlorine.	N by $KMnO_4$.	N as NH_4.	N as NO_2.	N as NO_3.
Rain water.	Bellefonte—collected by Prof. Wm. Frear, after long rain.	5	none	0.148	0.280	none	none
Subsoil water.	Wynnewood—pool fed by underground spring.	65	6.20	0.032	0.024	none	2.3
" "	Wynnewood—well about 150 yards from above.	60	4.00	0.024	0.016	none	3.5
" "	Wynnewood—well polluted by farm-yard drainage; about 500 yards from pool.	...	16.00	0.208	0.028	none	14.2
" "	Pump well in densely populated district. Highly contaminated.	1120	57.00	1.00	3 120	0.01	33.0
" "	Newly dug well in populated district. Highly contaminated.	620	120.00	0.08	2.00	0.03	16.0
" "	Well at Barren Hill, 130 feet deep.	470	120.00	undet.	undet.	traces	22.0

RESULTS FROM SCHUYLKILL RIVER—PARTS PER MILLION.

Samples taken from hydrant at 715 Walnut Street.

1888.	State of Weather.	Condition.	Total Solids.	Nitrogen by KMnO$_4$.	Nitrogen, as NH$_4$.	Nitrogen as NO$_2$.	Nitrogen as NO$_3$.
Sept. 17.	Continued rain.	Turbid.	160	.06	.048	None.	.13
" 18.	Continued rain.	Muddy.	160	.09	.020	"	.34
" 19.	Clear.	Turbid.	140	.108	.004	Trace.	.34
" 20.	"	"	150	.124	None.	Faint trace.	.25
" 21.	"	Less turbid.	150	.132	"	None.	.25
" 22.	"	Very muddy.	180	.180	"	Faint trace.	.25
" 24.	"	Turbid.	140	.100	"	None.	.13
" 25.	"	Slightly turbid.	130	.092	"	"	.13
" 26.	"	Slightly turbid.	120	.068	.01	"	.18
" 27.	"	Very slightly turbid.	110	.056	None.	"	.18
" 28.	"	Very slightly turbid.	105	.052	"	"	.36
" 29.	"	Nearly clear.	125	.052	.008	"	.70
Oct. 1.	"	"	140	.052	.012	"	.70
" 2.	"	"	125	.044	None.	"	.34
" 3.	"	Clear.	120	.083	.016	"	.40
" 4.	"	"	120	.084	.008	"	.68
" 5.	Rain.	"	120	.092	.012	"	.80
" 6.	"	Turbid.	120	.060	.012	"	.70
" 8.	Clear.	Slightly turbid.	150	.108	.008	"	.80
" 9.	"	Almost clear.	130	.088	.008	"	.70
" 10.	"	Clear.	125	.056	.010	"	.70
" 11.	"	"	115	.044	.010	"	.70

ANALYSES OF ARTESIAN WATERS.

MILLIGRAMS PER LITER.

	PHILADELPHIA.					BALTIMORE.	INDIANA.
	9th and Chestnut Sts.	24th and Washington Ave.	15th and Walnut Sts.	Broad and Columbia Ave.	Broad and Chestnut Sts.	Black's Island.	Locust Point.*	Ft. Wayne.
Condition,	Clear	Clear	Slightly turbid	Clear	Clear	Clear	Clear	Clear
Reaction,	Alkaline	Alkaline	Alkaline	Alkaline	Alkaline	Alkaline	Alkaline	Alkaline
SiO_2,	36.00	34.50	35.00	24.00	29.00	19.50	4.00	...
SO_4,	61.80	36.26	9.90	25.84	15.24	102.36	4.10	...
PO_4,	0.24	0.31	0.31	Trace	Trace	0.38	...	...
CO_3 (combined),	64.91	110.17	54.80	78.84	108.64	50.64	...	190.2
Cl,	89.21	91.77	108.50	206.01	172.00	554.48	4.68	...
H_2S,	...	Trace	...	...	...	...	...	...
Mn,	...	...	...	1.80	4.32	...	...	...
Fe,	9.80	7.70	5.70	5.28	...	1.05	2.5	...
Ca,	54.57	44.28	50.00	125.00	101.42	53.57	0.86	...
Mg,	15.27	7.98	12.60	30.70	22.27	6.48	...	...
Na,	54.13	84.13	34.46	72.34	55.83	378.63	...	...
K,	4.14	8.54	4.10	Traces	6.28	Trace	...	...
N by $KMnO_4$,	0.032	Not det.	0.048	0.048	0.015	0.148	.012	0.035
" as NH_4,	0.248	Not det.	0.034	0.032	0.004	0.132	.044	0.315
" " NO_2,	Trace	None	None	0.02	Trace	None	None.	None.
" " NO_3,	Trace	Trace	None	2.00	1.00	None	None.	None;

*We are indebted to Mr. Wm. Glenn, Chemist of the Baltimore Chrome Works, for this sample.

TABLE SHOWING THE RELATIVE PURITY OF THE WATER SUPPLIED TO CITIES, FROM THE DETERMINATIONS MADE IN JUNE, 1881, BY A. R. LEEDS.

Parts per million.	Philadelphia.	New York.	Brooklyn.	Jersey City.	Boston.	Washington.	Rochester.	Cincinnatti.
Total solids, . . .	143.0	118.0	60.0	93 0	85.0	115.0	100.0	162.0
" hardness, . .	44.0	33.0	22.7	32.0	21.0	48.0	55.0	64.0
Chlorine, ...	3.0	3.5	5.5	2.35	3.15	2.70	1.95	8.05
Oxygen-consuming power, ...	4 6	8.1	4.13	9.5	17.7	6.00	7.9	8.6
Nitrogen by KMnO$_4$, ...	.147	.221	.067	.344	.496	.221	.188	.196
Nitrogen in ammonium compounds.	.008	.022	.006	.039	.108	.049	.093	.094
Nitrogen as nitrites.	None	None	None	None	None	None	None	None
" " nitrates.	1.51	1.84	2.69	2.01	2.75	1.84	1.39	1.64

INDEX.

ACID phenyl sulphate, 28
——— Acids, action on lead, 79
Actinic method for organic matter, 37
Action of water on lead, 78
Aëration of water, 76
Agnew, D. H., on urinary calculi, 68
Albuminoid ammonia, 27, 72
Alkali carbonates, determination of, 59
Alkaline permanganate, 26
Allen, on boiler waters, 90-93
———, lead in water, 79
———, sulphuric acid in water, 90
———, test for zinc, 45
Alum, action of, 84
———, use of, 84
Aluminum, determination of, 50
——— in scale, 91
Ammonia, albuminoid, 27
———, free, 26
———, free water, 25
——— from rain water, 72
——— process, 23
Ammonium chloride, standard, 25
——— molybdate, 38
——— picrate solution, 29
Ammoniums, substitution, 28
Analysis, statement of, 66
Analytical operations, 16
Anderson and Ogsten, purification of water, 85
Animal matter, 73
Antwerp water, purification of, 85
Artesian water, 9, 14
——— waters, composition of, 101
Arsenic, detection of, 45
———, effect of, 69
Asbestos filters, 83

BACTERIA in water, 80
Bacteriological examination, 80
Barium, detection and estimation of, 44
Barren Hill well, 15, 99
Barus, Carl, suspended matters in water, 11

Basin platinum, 19
Bicarbonates in water, 93
Black's Island well, 69, 101
Blarez' oxygen process, 40
Boiler mud, 91
——— water, 89
——— water, points to be determined in, 92
——— ———, purification of, 93
——— ———, statement of results from, 92
Boric acid estimation, 60
Burner, low temperature, 20

CALCIUM bicarbonate, 93
——— carbonate, action in boiler waters, 91
——— ———, solubility of, 93
——— compounds, removal of, 93
———, determination of, 50, 51
——— hydroxide for purifying water, 94
——— sulphate, insolubility of, 92
——— sulphate, action in boiler water, 91, 92
Caramel solution, 43
Carbonates, action on lead, 79
———, determination of normal, 58
Carbon filters, 83
Carbonic acid, action of, 14
——— ———, free, determination of, 60
——— ———, effect on microbes, 82
Carleton-Williams, lead in water, 79
Carnelly and Frew, lead in water, 78, 79
Caustic soda, use of, in boiler water, 93
Cellar waters, examination of, 87
Chalk waters, 75
Chlorine, determination of, 22
———, significance of, 70
Chromium, detection of, 44
City supplies, 102
Clarifying water, 17, 84
Clark's process for purifying water, 94

104 INDEX.

Coke filter, 86
Color, determination of, 18
———, significance of, 68
Comparison cylinders, 25
Control determination, 21, 52
Conversion of ratios, 97
Cooper, A. J., delicacy of tests, 49
Copper, detection of, 49
———, effects of, 69
——— sulphate, standard, 49
Corrosion of boilers, 89
Crookes, Odling and Tidy, lead in water, 78
Cultivation of microbes, 81
Culture fluid, 81

DEEP water, 14, 70, 71, 74, 101
Demijohn for water samples, 16
Dibdin, table of dissolved oxygen, 98
Distilled water, wholesomeness of, 69, 83
Disease, transmission of, by water, 80
Driffield, composition of boiler mud, 91
Dupré, dissolved oxygen, 44, 76

EARTHENWARE filters, 83

FERMENTATION, butyric, 71
Ferrous ammonium sulphate, standard, 40
Ferrous compounds, action of, 34
Ferric sulphate, standard, 46
Filter beds, 84
——— paper, 17
Filters, efficacy of various forms of, 83
Filtration, effect of, 82, 83
Fleck's silver method, 37
Fluorescein, use of, 87
Forceps, platinum, 20
Fox's rate of evolution of ammonia, 72
Frankland, filtration, 83
———, purification by precipitation, 84
——— and Tidy, standards of purity, 76
——— sewage fungus, 39
Free acid, effects of, 82, 90
——— ammonia, 26, 71
Frew and Carnelly, lead in water, 78, 79
Funnel, safety, 25

GALLON, Imperial, 66
———, U. S., 66
Gelatin culture fluid, 81
———, liquefaction of, 81
Gérardin, dissolved oxygen, 76
Germs in water, 80
Gooch, method for borates, 60
———, method for lithium, 55

HARDNESS, determination of, 58
———, permanent, 59

Hardness, temporary, 59, 77, 92
Hard scale, 91
——— water, sanitary relations of, 68
——— water, softening of, 94, 95
Hehner, limit of phosphates, 70
Hehner's cylinders for color comparison, 25
——— method for hardness, 58
Heisch's test, 38, 71
History of water, 9
Hunt, T. Sterry, water in rocks, 12
Hydrochloric acid, diluted, 31
Hydrogen sulphide, titration of, 56, 57

IDENTIFICATION of source of water, 86
Imperial gallon, 66
Interpretation of results, 66
Iodine, centinormal, 57
Iron, action of, in filtration, 85
——— compounds, solution by water, 14
———, determination of, 46, 50
———, granular, for purification, 85
———, in scale, 91
———, significance of, 69

JUDGMENT on analytical data, 67

KOCH'S culture method, 80

LACMOID, use of, 19, 58
Lead, action of water on, 78
Lead, determination of, 48
———, nitrate, standard, 48
———, significance of, 69
Leeds, actinic method, 37
———, dissolved oxygen, 76
Lime, purification of water by, 94
Lithium compounds, use of, 86
———, detection of, 63
———, separation of, 53, 55
Litmus, use of, 19
Locust Point well, 15, 101
Lott, F. E , Heisch's test, 71

MAGNESIA in boiler sludge, 91
Magnesium, determination of, 50, 51
Magnesium chloride, decompositions and effects of, 90
——— compounds, removal of, 93, 95, 96
Mallet, ammonia process, 73
Manganese, detection of, 47
———, determination of, 50, 52
Marsh's test, 45
Microbes, action of, 12, 13, 81
———, pathogenic, 82
Mineral springs, 14
Mine water, 90
Müller, lead in water, 79

INDEX.

NAPHTHYLAMINE hydrochloride, 30
Naphthylammonium chloride, 30
Nesslerizing, 26
Nessler's reagent, 25
Nickel dish, 20, 59
Nitric acid, diluted, 47
Nitrification, 13
Nitrates, action in boilers, 91
——, determination of, 28
——, formation of, 13
——, significance of, 75
Nitrites, action of, 34
——, determination of, 30
——, formation of, 13
——, removal of, 34
——, significance of, 74
Nitrogen in ammonium compounds, 23, 71
—— as nitrites, 30
—— as nitrates, 28
——, oxidation of, 13
—— by permanganate, 72

ODOR, determination of, 18
—— from residue, 21
——, significance of, 68, 69
Odling, Crookes and Tidy, lead in water, 78
Organic matter, 12, 21, 67, 77
—— ——, action of, 14
—— ——, nature of, 28, 32, 73
—— ——, oxidation of, 32
—— ——, removal of, 85
Organisms, detection of, 77
—— in water, 80
——, precipitation of, 84
Oxygen consumed, 33
—— consuming microbes, 44
—— -consuming power, 32, 73, 75
——, amount of, dissolved, 98
——, dissolved, determination of, 40
——, dissolved, effects of, 76, 89

PAPER filters, 83
Para - amido - benzene - sulphonic acid, 30
Pasteur's filters, 83
Pathogenic organisms, 81
Permanganate method, 35, 36
—— standard, 33, 40
Petroleum 'in water, 89
Pettenkofer's method for free carbonic acid, 60
Phenacetolin, use of, 58
Phenolphthaleïn, use of, 19
Phosphates, action on lead, 79
——, determination of, 38, 50
——, significance of, 70
——, use of, in purifying water, 95
Phenyl sulphuric acid, 28
Picric acid, 29

J

Plagge, filtration, 83
Plate culture, 80
Platinum, preservation of, 20
Plummet for specific gravity, 64
Poisonous metals, detection of, 44
—— ——, significance of, 69
Polluted waters, characters of, 67–77
Porter's process for purifying water, 94
Potassium, determination, 52
—— chromate solution, 22
—— iodide solution, 35
—— nitrate, standard, 28
Potassium permanganate, alkaline, 26
—— permanganate, decinormal, 40
Preliminary examination, 16
Ptomaïnes, 12
Pure water, corrosive action of, 89
Purification of boiler water, 95
—— of drinking waters, 83
Purity, standards of, 77
Pyrogallol method for oxygen, 43
Pyrogallol solution, 43

QUANTITATIVE analysis, 50

RAIN water, 9, 72, 75
—— ——, composition of, 99
Ratios, conversion of, 97
Reaction, 19
Residue, charring of, 21
Results, statement of, 66
River water, 9, 10, 75, 100

SALT, action on boiler waters, 90
Salamon and Matthews, coke filters, 86
Samples, collection of, 16
Sand filters, 83
Sanitary application, 67
—— examinations, 16
Scale, 91
Schuylkill River water, composition of, 10, 100
Sewage, action of, 14, 75
—— fungus, 38, 39
Sewer water, detection of, 88
Silica, action of, in water, 78
——, determination of, 50
—— in scale, 91
Silicates, action on lead, 79
Silver nitrate, standard, 22
—— ——, test in Porter's process, 94
—— nitrite, preparation of, 31
—— test for organic matter, 37
Sludge, 91
Solids total, determination of, 19
Solids, significance of, 68
Sodium and potassium, separation of, 54
—— carbonate, solution of, 25
—— ——, standard, 58
—— ——, use of in boiler waters, 94

INDEX.

Sodium chloride, standard, 22
——, determination of, 52
—— hydroxide, solution, 40, 43
—— ——, use in boiler waters, 94
—— nitrite, standard, 31
—— thiosulphate, 35
Source of water, tracing of, 86
Smart, C., nature of organic matter, 73
Subsoil waters, composition of, 12, 99
Specific gravity, 64
—— —— bottle, 64
Spectroscope, 63
Spectroscopic examination, 62
Spongy iron filters, 83
—— ——, use of, 85
Standards of purity, 77
Starch indicator, 35
Stone filters, 83
Storage, effect of, 81
Strontium compounds, use of, 86
Subsidence, promotion of, 17, 84
Subsoil water, 9, 11, 75, 99
Sugar test, 38
Sulphanilic acid, 30
Sulphur compounds, action of, 34
Sulphides, formation of, 14
Sulphuretted hydrogen, determination of, 14, 56
Sulphuric acid, diluted, 33
—— ——, standard, 58
Surface water, composition of, 9, 10, 100
Suspended matters, 10, 11

TASTE, significance of, 68
Technical examinations, 50
Tests for metals, delicacy of, 49
Thompson, estimation of iron, 46
Tidy's permanganate process, 35
Tidy and Frankland, standard of purity, 76
——, Odling and Crookes, lead in water, 78
Tri-sodium phosphate, use of, for purification, 95
Trinitrophenol, 29
Turquem's burner, 62

UNCONTAMINATED waters, 77
Urea, decomposition of, 72
Urine in water, 72
U. S. gallon, 66

VEGETABLE matter, 21, 73
Vegetable growth in water, 76

WARINGTON, R., nitrification, 12
Water, amount of, in rocks, 12
Wanklyn, standards of purity, 74
Wanklyn's albuminoid ammonia process, 23
—— test for manganese, 47

ZINC, detection of, 45
——, effect of, 69

CATALOGUE No. 7.   MARCH, 1889.

A CATALOGUE

OF

BOOKS FOR STUDENTS.

INCLUDING THE

? QUIZ-COMPENDS ?

CONTENTS.

	PAGE		PAGE
New Series of Manuals,	2,3,4,5	Obstetrics.	. 10
Anatomy,	. 6	Pathology, Histology,	. 11
Biology,	. 11	Pharmacy,	13
Chemistry,	. 6	Physical Diagnosis,	11
Children's Diseases,	7	Physiology,	11
Dentistry,	. 8	Practice of Medicine,	. 12
Dictionaries,	. 8	Prescription Books,	. 12
Eye Diseases,	8	? Quiz-Compends ?	15, 16
Electricity,	9	Skin Diseases,	. 13
Gynæcology,	. 10	Surgery,	. 13
Hygiene,	9	Therapeutics,	. 9
Materia Medica,	. 9	Throat,	. 14
Medical Jurisprudence,	. 9	Urine and Urinary Organs,	14
Miscellaneous,	10	Venereal Diseases,	. 14

PUBLISHED BY

P. BLAKISTON, SON & CO.,

Medical Booksellers, Importers and Publishers.

LARGE STOCK OF ALL STUDENTS' BOOKS, AT
THE LOWEST PRICES.

1012 Walnut Street, Philadelphia.

₄ For sale by all Booksellers, or any book will be sent by mail, postpaid, upon receipt of price. Catalogues of books on all branches of Medicine, Dentistry, Pharmacy, etc., supplied upon application.

"*An excellent Series of Manuals.*"—*Archives of Gynæcology*

A NEW SERIES OF
STUDENTS' MANUALS

On the various Branches of Medicine and Surgery.

Can be used by Students of any College.

Price of each, Handsome Cloth, $3.00. Full Leather, $3.50.

The object of this series is to furnish good manuals for the medical student, that will strike the medium between the compend on one hand and the prolix text-book on the other—to contain all that is necessary for the student, without embarrassing him with a flood of theory and involved statements. They have been prepared by well-known men, who have had large experience as teachers and writers, and who are, therefore, well informed as to the needs of the student.

Their mechanical execution is of the best—good type and paper, handsomely illustrated whenever illustrations are of use, and strongly bound in uniform style.

Each book is sold separately at a remarkably low price, and the immediate success of several of the volumes shows that the series has met with popular favor.

No. 1. SURGERY. 236 Illustrations.

A Manual of the Practice of Surgery. By WM. J. WALSHAM, M.D., Asst. Surg. to, and Demonstrator of Surg. in, St. Bartholomew's Hospital, London, etc. 228 Illustrations.

Presents the introductory facts in Surgery in clear, precise language, and contains all the latest advances in Pathology, Antiseptics, etc.

"It aims to occupy a position midway between the pretentious manual and the cumbersome System of Surgery, and its general character may be summed up in one word—practical."—*The Medical Bulletin.*

"Walsham, besides being an excellent surgeon, is a teacher in its best sense, and having had very great experience in the preparation of candidates for examination, and their subsequent professional career, may be relied upon to have carried out his work successfully. Without following out in detail his arrangement, which is excellent, we can at once say that his book is an embodiment of modern ideas neatly strung together, with an amount of careful organization well suited to the candidate, and, indeed, to the practitioner."—*British Medical Journal.*

Price of each Book, Cloth, $3.00; Leather, $3.50.

No. 2. DISEASES OF WOMEN. 130 Illus.
The Diseases of Women. By DR. F. WINCKEL, Professor of Gynæcology and Director of the Royal University Clinic for Women, in Munich. Translated from the German by DR. J. H. WILLIAMSON, Resident Physician Allegheny General Hospital, Allegheny, Penn'a, under the supervision of, and with an introduction by, **Theophilus Parvin**, M.D., Professor of Obstetrics and Diseases of Women and Children in Jefferson Medical College. Illustrated by 132 fine Engravings on Wood, most of which are new.

"The book will be a valuable one to physicians, and a safe and satisfactory one to put into the hands of students. It is issued in a neat and attractive form, and at a very reasonable price."—*Boston Medical and Surgl. Journal.*

No. 3. OBSTETRICS. 227 Illustrations.
A Manual of Midwifery. By ALFRED LEWIS GALABIN, M.A., M.D., Obstetric Physician and Lecturer on Midwifery and the Diseases of Women at Guy's Hospital, London; Examiner in Midwifery to the Conjoint Examining Board of England, etc. With 227 Illus.

"This manual is one we can strongly recommend to all who desire to study the science as well as the practice of midwifery. Students at the present time not only are expected to know the principles of diagnosis, and the treatment of the various emergencies and complications that occur in the practice of midwifery, but find that the tendency is for examiners to ask more questions relating to the science of the subject than was the custom a few years ago. * * * The general standard of the manual is high; and wherever the science and practice of midwifery are well taught it will be regarded as one of the most important text-books on the subject."—*London Practitioner.*

No. 4. PHYSIOLOGY. Third Edition.
321 ILLUSTRATIONS AND A GLOSSARY.
A Manual of Physiology. By GERALD F. YEO, M.D., F.R.C.S., Professor of Physiology in King's College, London. 321 Illustrations and a Glossary of Terms. Third American from second English Edition, revised and improved. 758 pages.

This volume was specially prepared to furnish students with a new text-book of Physiology, elementary so far as to avoid theories which have not borne the test of time and such details of methods as are unnecessary for students in our medical colleges.

"The brief examination I have given it was so favorable that I placed it in the list of text-books recommended in the circular of the University Medical College."—*Prof. Lewis A. Stimson*, M.D., *37 East 33d Street, New York.*

Price of each Book, Cloth, $3.00; Leather, $3.50.

No. 5. POTTER'S MATERIA MEDICA, PHARMACY AND THERAPEUTICS.

OVER 600 PRESCRIPTIONS, FORMULÆ, ETC.

A Handbook of Materia Medica, Pharmacy and Therapeutics—including the Physiological Action of Drugs, Special Therapeutics of Diseases, Official and Extemporaneous Pharmacy, etc., etc. By SAM'L O. L. POTTER, M.A., M.D., Professor of the Practice of Medicine in Cooper Medical College, San Francisco, Late A. A. Surg., U. S. A., Author of the "Quiz-Compends" of Anatomy and Materia Medica, etc.

This book contains many unique features of style and arrangement; no time or trouble has been spared to make it most complete and yet concise in all its parts. It contains 600 prescriptions of practical worth, a great mass of facts conveniently and concisely put together, also many tables, dose lists, diagnostic hints, etc., all rendering it the most complete manual ever published.

☞ Part III, Special Therapeutics, consists of an Alphabetical List of Diseases, in which is given the proper drugs to be used in the treatment of each, with the authority recommending them, and in many cases signed prescriptions. This will be found of great value to the young practitioner, and to the physician of experience it will suggest new methods of treatment in obstinate and chronic cases.

"Dr. Potter's handbook will find a place, and a very important one, in our colleges and the libraries of our practitioners."—*N. Y. Medical Journal.*

No. 6. DISEASES OF CHILDREN.

A Manual. By J. F. GOODHART, M.D., Phys. to the Evelina Hospital for Children; Asst. Phys. to Guy's Hospital, London. American Edition. Edited by LOUIS STARR, M.D., Clinical Prof. of Dis. of Children in the Hospital of the Univ. of Pennsylvania, and Physician to the Children's Hospital, Phila. Containing many new Prescriptions, a list of over 50 Formulæ, conforming to the U. S. Pharmacopœia, and Directions for making Artificial Human Milk, for the Artificial Digestion of Milk, etc.

"As it is said of some men, so it might be said of some books, that they are 'born to greatness.' This new volume has, we believe, a mission, particularly in the hands of the younger members of the profession. In these days of prolixity in medical literature, it is refreshing to meet with an author who knows both what to say and when he has said it. The work of Dr. Goodhart

Price of each Book, Cloth, $3.00 ; Leather, $3.50.

(admirably conformed, by Dr. Starr, to meet American requirements) is the nearest approach to clinical teaching without the actual presence of clinical material that we have yet seen."—*New York Medical Record.*

No. 7. PRACTICAL THERAPEUTICS.
FOURTH EDITION, WITH AN INDEX OF DISEASES.

Practical Therapeutics, considered with reference to Articles of the Materia Medica. Containing, also, an Index of Diseases, with a list of the Medicines applicable as Remedies. By EDWARD JOHN WARING, M.D., F.R.C.P. Fourth Edition. Rewritten and Revised. By DUDLEY W. BUXTON, M.D., Asst. to the Prof. of Medicine at University College Hospital.

"We wish a copy could be put in the hands of every Student or Practitioner in the country. In our estimation, it is the best book of the kind ever written."—*N. Y. Medical Journal.*

No. 8. MEDICAL JURISPRUDENCE AND TOXICOLOGY. New Ed.

By JOHN J. REESE, M.D., Professor of Medical Jurisprudence and Toxicology in the University of Pennsylvania; President of the Medical Jurisprudence Society of Phila.; 2d Edition, Revised and Enlarged.

"This admirable text-book."—*Amer. Jour. of Med. Sciences.*

"We lay this volume aside, after a careful perusal of its pages, with the profound impression that it should be in the hands of every doctor and lawyer. It fully meets the wants of all students. . . . . He has succeeded in admirably condensing into a handy volume all the essential points."—*Cincinnati Lancet and Clinic.*

No. 9. ORGANIC CHEMISTRY.

Or the Chemistry of the Carbon Compounds. By Prof. VICTOR VON RICHTER, University of Breslau. Authorized translation, from the Fourth German Edition. By EDGAR F. SMITH, M.A., PH.D.; Prof. of Chemistry in University of Pennsylvania; Member of the Chem. Socs. of Berlin and Paris.

"I must say that this standard treatise is here presented in a remarkably compendious shape."—*J. W. Holland,* M.D., *Professor of Chemistry, Jefferson Medical College, Philadelphia.*

"This work brings the whole matter, in simple, plain language, to the student in a clear, comprehensive manner. The whole method of the work is one that is more readily grasped than that of older and more famed text-books, and we look forward to the time when, to a great extent, this work will supersede others, on the score of its better adaptation to the wants of both teacher and student."—*Pharmaceutical Record.*

Price of each Book, Cloth, $3.00; Leather, $3.50.

ANATOMY.

Holden's Anatomy. A manual of Dissection of the Human Body. Fifth Edition. Enlarged, with Marginal References and over 200 Illustrations. Octavo. Cloth, 5.00; Leather, 6.00
Bound in Oilcloth, for the Dissecting Room, $4.50.

"No student of Anatomy can take up this book without being pleased and instructed. Its Diagrams are original, striking and suggestive, giving more at a glance than pages of text description. * * * The text matches the illustrations in directness of practical application and clearness of detail."—*New York Medical Record.*

Holden's Human Osteology. Comprising a Description of the Bones, with Colored Delineations of the Attachments of the Muscles. The General and Microscopical Structure of Bone and its Development. With Lithographic Plates and Numerous Illustrations. Seventh Edition. 8vo. Cloth, 6.00

Holden's Landmarks, Medical and Surgical. 4th ed.
Cloth, 1.25

Heath's Practical Anatomy. Sixth London Edition. 24 Colored Plates, and nearly 300 other Illustrations. Cloth, 5.00

Potter's Compend of Anatomy. Fourth Edition. 117 Illustrations. Cloth, 1.00; Interleaved for Notes, 1.25

CHEMISTRY.

Bartley's Medical Chemistry. A text-book prepared specially for Medical, Pharmaceutical and Dental Students. With 40 Illustrations, Plate of Absorption Spectra and Glossary of Chemical Terms. Cloth, 2.50

₊ This book has been written especially for students and physicians. It is practical and concise, dealing only with those parts of chemistry pertaining to medicine; no time being wasted in long descriptions of substances and theories of interest only to the advanced chemical student.

Bloxam's Chemistry, Inorganic and Organic, with Experiments. Sixth Edition. Enlarged and Rewritten. Nearly 300 Illustrations. Cloth, 4.50; Leather, 5.50

Richter's Inorganic Chemistry. A text-book for Students. Third American, from Fifth German Edition. Translated by Prof. Edgar F. Smith, PH.D. 89 Wood Engravings and Colored Plate of Spectra. Cloth, 2.00

Richter's Organic Chemistry, or Chemistry of the Carbon Compounds. Translated by Prof. Edgar F. Smith, PH.D. Illustrated. Cloth, 3.00; Leather, 3.50

☞ *See pages 2 to 5 for list of Students' Manuals.*

Chemistry:—Continued.

Trimble. Practical and Analytical Chemistry. A Course in Chemical Analysis, by Henry Trimble, Prof. of Analytical Chemistry in the Phila. College of Pharmacy. Illustrated. Second Edition. 8vo. Cloth, 1.50

Tidy. Modern Chemistry. 2d Ed. Cloth, 5.50

Leffmann's Compend of Chemistry. Inorganic and Organic. Including Urinary Analysis and the Sanitary Examination of Water. New Edition. Cloth, 1.00; Interleaved for Notes, 1.25

Muter. Practical and Analytical Chemistry. Second Edition, Revised and Illustrated. Cloth, 2.00

Holland. The Urine, and Common Poisons, Chemical and Microscopical. For Laboratory Use. 3d Edition, Enlarged. Illustrated. *In Press.*

Van Nüys. Urine Analysis. Illus. Cloth, 2.00

Wolff's Applied Medical Chemistry. By Lawrence Wolff, M.D., Demonstrator of Chemistry in Jefferson Medical College, Philadelphia. Cloth, 1.00

CHILDREN.

Goodhart and Starr. The Diseases of Children. A Manual for Students and Physicians. By J. F. Goodhart, M.D., Physician to the Evelina Hospital for Children; Assistant Physician to Guy's Hospital, London. American Edition, Revised and Edited by Louis Starr, M.D., Clinical Professor of Diseases of Children in the Hospital of the University of Pennsylvania; Physician to the Children's Hospital, Philadelphia. Containing many new Prescriptions, a List of over 50 Formulæ, conforming to the U. S. Pharmacopœia, and Directions for making Artificial Human Milk, for the Artificial Digestion of Milk, etc.
Cloth, 3.00; Leather, 3.50

Day. On Children. A Practical and Systematic Treatise. Second Edition. 8vo. 752 pages. Cloth, 3.00; Leather, 4.00

Meigs and Pepper. The Diseases of Children. Seventh Edition. 8vo. Cloth, 5.00; Leather, 6.00

Starr. Diseases of the Digestive Organs in Infancy and Childhood. With chapters on the Investigation of Disease, and on the General Management of Children. By Louis Starr, M.D., Clinical Professor of Diseases of Children in the University of Pennsylvania; with a section on Feeding, including special Diet Lists, etc. Illus. Cloth, 2.50

☞ *See pages 15 and 16 for list of ? Quiz-Compends ?*

DENTISTRY.

Fillebrown. Operative Dentistry. 330 Illustrations. *Just Ready.* Cloth, 2.50

Flagg's Plastics and Plastic Filling. 3d Ed. *Preparing.*

Gorgas. Dental Medicine. A Manual of Materia Medica and Therapeutics. Second Edition. Cloth, 3.25

Harris. Principles and Practice of Dentistry. Including Anatomy, Physiology, Pathology, Therapeutics, Dental Surgery and Mechanism. Twelfth Edition. Revised and enlarged by Professor Gorgas. 1028 Illustrations. Cloth, 7.00; Leather, 8.00

Richardson's Mechanical Dentistry. Fifth Edition. 569 Illustrations. 8vo. Cloth, 4.50; Leather, 5.50

Stocken's Dental Materia Medica. Third Edition. Cloth, 2.50

Taft's Operative Dentistry. Dental Students and Practitioners. Fourth Edition. 100 Illustrations. Cloth, 4.25; Leather, 5.00

Talbot. Irregularities of the Teeth, and their Treatment. Illustrated. 8vo. Cloth, 2.00

Tomes' Dental Anatomy. Third Ed. 191 Illus. *Preparing.*

Tomes' Dental Surgery. Third Edition. Revised. 292 Illustrations. 772 Pages. Cloth, 5.00

DICTIONARIES.

Cleaveland's Pocket Medical Lexicon. Thirty-first Edition. Giving correct Pronunciation and Definition of Terms used in Medicine and the Collateral Sciences. Very small pocket size.
Cloth, red edges .75; pocket-book style, 1.00

Longley's Pocket Dictionary. The Student's Medical Lexicon, giving Definition and Pronunciation of all Terms used in Medicine, with an Appendix giving Poisons and Their Antidotes, Abbreviations used in Prescriptions, Metric Scale of Doses, etc. 24mo. Cloth, 1.00; pocket-book style, 1.25

EYE.

Arlt. Diseases of the Eye. Including those of the Conjunctiva, Cornea, Sclerotic, Iris and Ciliary Body. By Prof. Von Arlt. Translated by Dr. Lyman Ware. Illus. 8vo. Cloth, 2.50

Hartridge on Refraction. 3d Ed. Cloth, 2.00

Macnamara. Diseases of the Eye. 4th Edition. Revised. Colored Plates and Wood Cuts and Test Types. Cloth, 4.00

Meyer. Diseases of the Eye. A complete Manual for Students and Physicians. 270 Illustrations and two Colored Plates. 8vo. Cloth, 4.50; Leather, 5.50

Fox and Gould. Compend of Diseases of the Eye and Refraction. 2d Ed. Enlarged. 71 Illus. 39 Formulæ.
Cloth, 1.00; Interleaved for Notes, 1.25

☞ *See pages 2 to 5 for list of Students' Manuals.*

ELECTRICITY.

Mason's Compend of Medical and Surgical Electricity.
With numerous Illustrations. 12mo. Cloth, 1.00

HYGIENE.

Parkes' Practical Hygiene. Seventh Edition, enlarged. Illustrated. 8vo. Cloth, 4.50
Wilson's Handbook of Hygiene and Sanitary Science. Sixth Edition. Revised and Illustrated. Cloth, 2.75

MATERIA MEDICA AND THERAPEUTICS.

Potter's Compend of Materia Medica, Therapeutics and Prescription Writing. Fifth Edition, revised and improved.
Cloth, 1.00; Interleaved for Notes, 1.25
Biddle's Materia Medica. Eleventh Edition. By the late John B. Biddle, M.D., Professor of Materia Medica in Jefferson Medical College, Philadelphia. Thoroughly revised, and in many parts rewritten, by his son, Clement Biddle, M.D., Assistant Surgeon, U. S. Navy, assisted by Henry Morris, M.D., Demonstrator of Obstetrics in Jefferson Medical College. 8vo., illustrated. Cloth, 4.00; Leather, 4.75
Headland's Action of Medicines. 9th Ed. 8vo. Cloth, 3.00
Potter. Materia Medica, Pharmacy and Therapeutica. Including Action of Medicines, Special Therapeutics, Pharmacology, etc. *Page 4.* Cloth, 3.00; Leather, 3.50
Starr, Walker and Powell. Synopsis of Physiological Action of Medicines, based upon Prof. H. C. Wood's "Materia Medica and Therapeutics." 3d Ed Enlarged. Cloth, .75
Waring. Therapeutics. With an Index of Diseases and an Index of Remedies. A Practical Manual. Fourth Edition. Revised and Enlarged. Cloth, 3.00; Leather, 3.50

MEDICAL JURISPRUDENCE.

Reese. A Text-book of Medical Jurisprudence and Toxicology. By John J. Reese, M.D., Professor of Medical Jurisprudence and Toxicology in the Medical Department of the University of Pennsylvania; President of the Medical Jurisprudence Society of Philadelphia; Physician to St. Joseph's Hospital; Corresponding Member of The New York Medico-legal Society. 2d Edition. Cloth, 3.00; Leather, 3.50
Woodman and Tidy's Medical Jurisprudence and Toxicology. Chromo-Lithographic Plates and 116 Wood engravings.
Cloth, 7.50; Leather, 8.50

☞ *See pages 15 and 16 for list of ? Quiz-Compends ?*

MISCELLANEOUS.

Allingham. Diseases of the Rectum. Fourth Edition. Illustrated. 8vo. Paper covers, .75; Cloth, 1.25

Beale. Slight Ailments. Their Nature and Treatment. Illustrated. . 8vo. Paper cover, .75; Cloth, 1.25

Domville on Nursing. 6th Edition. Cloth, .75

Fothergill. Diseases of the Heart, and Their Treatment. Second Edition. 8vo. Cloth, 3.50

Gowers. Diseases of the Nervous System. 341 Illustrations. Cloth, 6.50; Leather, 7.50

Mann's Manual of Psychological Medicine, and Allied Nervous Diseases. Their Diagnosis, Pathology and Treatment, and their Medico-Legal Aspects. Illus. Cloth, 5.00; Leather, 6.00

Tanner. Memoranda of Poisons. Their Antidotes and Tests. Sixth Edition. Revised by Henry Leffmann, M.D. Cloth, .75

OBSTETRICS AND GYNÆCOLOGY.

Byford. Diseases of Women. The Practice of Medicine and Surgery, as applied to the Diseases and Accidents Incident to Women. By W. H. Byford, A.M., M.D., Professor of Gynæcology in Rush Medical College and of Obstetrics in the Woman's Medical College, etc., and Henry T. Byford, M.D., Surgeon to the Woman's Hospital of Chicago; Gynæcologist to St Luke's Hospital, etc. Fourth Edition. Revised, Rewritten and Enlarged. With 306 Illustrations, over 100 of which are original. Octavo 832 pages. Cloth, 5.00; Leather, 6.00

Cazeaux and Tarnier's Midwifery. With Appendix, by Mundé. The Theory and Practice of Obstetrics; including the Diseases of Pregnancy and Parturition, Obstetrical Operations, etc. By P. Cazeaux. Remodeled and rearranged, with revisions and additions, by S. Tarnier, M.D., Professor of Obstetrics and Diseases of Women and Children in the Faculty of Medicine of Paris. Eighth American, from the Eighth French and First Italian Edition. Edited by Robert J. Hess, M.D., Physician to the Northern Dispensary, Philadelphia, with an appendix by Paul F. Mundé, M.D., Professor of Gynæcology at the N. Y. Polyclinic. Illustrated by Chromo-Lithographs, Lithographs, and other Full-page Plates, seven of which are beautifully colored, and numerous Wood Engravings. *Students' Edition.* One Vol., 8vo. Cloth, 5.00; Leather, 6.00

Lewers' Diseases of Women. A Practical Text-Book. 139 Illustrations. Cloth, 2.25

Parvin's Winckel's Diseases of Women. Edited by Prof. Theophilus Parvin, Jefferson Medical College, Philadelphia. 117 Illustrations. *See page 3.* Cloth, 3.00; Leather, 3.50

Morris. Compend of Gynæcology. Illustrated. *In Press.*

☞ *See pages 2 to 5 for list of New Manuals.*

STUDENTS' TEXT-BOOKS AND MANUALS. 11

Obstetrics and Gynæcology:—Continued.
Landis' Compend of Obstetrics. Illustrated. 4th edition, enlarged. Cloth, 1.00; Interleaved for Notes, 1.25
Galabin's Midwifery. A New Manual for Students. By A. Lewis Galabin, M.D., F.R.C.P., Obstetric Physician to Guy's Hospital, London, and Professor of Obstetrics in the same Institution. 227 Illustrations. *See page 3.* Cloth, 3.00; Leather, 3.50
Glisan's Modern Midwifery. 2d Edition. Cloth, 3.00
Rigby's Obstetric Memoranda. By Alfred Meadows, M.D. 4th Edition. Cloth, .50
Meadows' Manual of Midwifery. Including the Signs and Symptoms of Pregnancy, Obstetric Operations, Diseases of the Puerperal State, etc. 145 Illustrations. 494 pages. Cloth, 2.00
Swayne's Obstetric Aphorisms. For the use of Students commencing Midwifery Practice. 8th Ed. 12mo. Cloth, 1.25

PATHOLOGY. HISTOLOGY. BIOLOGY.

Bowlby. Surgical Pathology and Morbid Anatomy, for Students. 135 Illustrations. 12mo. Cloth, 2.00
Davis' Elementary Biology. Illustrated. Cloth, 4.00
Rindfleisch's General Pathology. By Prof. Edward Rindfleisch. Translated by Wm. H. Mercur, M.D. Edited by James Tyson, M.D., Professor of Clinical Medicine in the University of Pennsylvania. 12mo. Cloth, 2.00
Gilliam's Essentials of Pathology. A Handbook for Students. 47 Illustrations. 12mo. Cloth, 2.00

** The object of this book is to unfold to the beginner the fundamentals of pathology in a plain, practical way, and by bringing them within easy comprehension to increase his interest in the study of the subject.

Gibbes' Practical Histology and Pathology. Third Edition. Enlarged. 12mo. Cloth, 1.75
Virchow's Post-Mortem Examinations. 2d Ed. Cloth, 1.00

PHYSICAL DIAGNOSIS.

Bruen's Physical Diagnosis of the Heart and Lungs. By Dr. Edward T. Bruen, Assistant Professor of Clinical Medicine in the University of Pennsylvania. Second Edition, revised. With new Illustrations. 12mo. Cloth, 1.50

PHYSIOLOGY.

Yeo's Physiology. Third Edition. The most Popular Students' Book. By Gerald F. Yeo, M.D., F.R.C.S., Professor of Physiology in King's College, London. Small Octavo. 758 pages. 321 carefully printed Illustrations. With a Full Glossary and Index. *See Page 3.* Cloth, 3.00; Leather, 3.50

☞ *See pages 15 and 16 for list of ? Quiz-Compends?*

12 STUDENTS' TEXT-BOOKS AND MANUALS.

Physiology:—Continued.

Brubaker's Compend of Physiology. Illustrated. Fourth Edition.   Cloth, 1.00; Interleaved for Notes, 1.25

Stirling. Practical Physiology, including Chemical and Experimental Physiology. 142 Illustrations.   Cloth, 2.25

Kirke's Physiology. New 12th Ed. Thoroughly Revised and Enlarged. 502 Illustrations.   Cloth, 4.00; Leather, 5.00

Landois' Human Physiology. Including Histology and Microscopical Anatomy, and with special reference to Practical Medicine. Second Edition. Translated and Edited by Prof. Stirling. 583 Illustrations.   Cloth, 6.50; Leather, 7.50

"So great are the advantages offered by Prof. Landois' Textbook, from the exhaustive and eminently practical manner in which the subject is treated, that, notwithstanding it is one of the largest works on Physiology, it has yet passed through four large editions in the same number of years. Dr. Stirling's annotations have materially added to the value of the work. . . . Admirably adapted for the practitioner. . . . With this Text-book at his command, no student could fail in his examination."—*Lancet.*

Sanderson's Physiological Laboratory. Being Practical Exercises for the Student. 350 Illustrations. 8vo.   Cloth, 5.00

Tyson's Cell Doctrine. Its History and Present State. Illustrated. Second Edition.   Cloth, 2.00

PRACTICE.

Roberts' Practice. New Revised Edition. A Handbook of the Theory and Practice of Medicine. By Frederick T. Roberts, M.D.; M.R.C.P., Professor of Clinical Medicine and Therapeutics in University College Hospital, London. Seventh Edition. Octavo.   Cloth, 5.50; Sheep, 6.50

Hughes. Compend of the Practice of Medicine. 3d Ed. Two parts, each, Cloth, 1.00; Interleaved for Notes, 1.25

    PART I.—Continued, Eruptive and Periodical Fevers, Diseases of the Stomach, Intestines, Peritoneum, Biliary Passages, Liver, Kidneys, etc., and General Diseases, etc.

    PART II.—Diseases of the Respiratory System, Circulatory System and Nervous System; Diseases of the Blood, etc.

Tanner's Index of Diseases, and Their Treatment. Cloth, 3.00

"This work has won for itself a reputation. . . . It is, in truth, what its Title indicates."—*N. Y. Medical Record.*

PRESCRIPTION BOOKS.

Wythe's Dose and Symptom Book. Containing the Doses and Uses of all the principal Articles of the Materia Medica, etc. Seventeenth Edition. Completely Revised and Rewritten. *Just Ready.* 32mo.   Cloth, 1.00; Pocket-book style, 1.25

Pereira's Physician's Prescription Book. Containing Lists of Terms, Phrases, Contractions and Abbreviations used in Prescriptions, Explanatory Notes, Grammatical Construction of Prescriptions, etc., etc. By Professor Jonathan Pereira, M.D. Sixteenth Edition. 32mo. Cloth, 1.00; Pocket-book style, 1.25

☞ *See pages 2 to 5 for list of New Manuals.*

STUDENTS' TEXT-BOOKS AND MANUALS. 13

PHARMACY.

Stewart's Compend of Pharmacy. Based upon Remington's Text-Book of Pharmacy. Second Edition, Revised.
Cloth, 1.00; Interleaved for Notes, 1.25

SKIN DISEASES.

Anderson, (McCall) Skin Diseases. A complete Text-Book, with Colored Plates and numerous Wood Engravings. 8vo. *Just Ready.* Cloth, 4.50; Leather, 5.50

"We welcome Dr. Anderson's work not only as a friend, but as a benefactor to the profession, because the author has stricken off mediæval shackles of insuperable nomenclature and made crooked ways straight in the diagnosis and treatment of this hitherto but little understood class of diseases. The chapter on Eczema is alone worth the price of the book."—*Nashville Medical News.*

"Worthy its distinguished author in every respect; a work whose practical value commends it not only to the practitioner and student of medicine, but also to the dermatologist."—*James Nevens Hyde*, M.D., *Prof. of Skin and Venereal Diseases, Rush Medical College, Chicago.*

Van Harlingen on Skin Diseases. A Handbook of the Diseases of the Skin, their Diagnosis and Treatment (arranged alphabetically). By Arthur Van Harlingen, M.D., Clinical Lecturer on Dermatology, Jefferson Medical College; Prof. of Diseases of the Skin in the Philadelphia Polyclinic. 2d Edition. Enlarged. With colored and other plates and illustrations. 12mo. Cloth, 2.50

Bulkley. The Skin in Health and Disease. By L. Duncan Bulkley, Physician to the N. Y. Hospital. Illus. Cloth, .50

SURGERY.

Heath's Minor Surgery, and Bandaging. Eighth Edition. 142 Illustrations. 60 Formulæ and Diet Lists. Cloth, 2.00

Horwitz's Compend of Surgery, including Minor Surgery, Amputations, Fractures, Dislocations, Surgical Diseases, and the Latest Antiseptic Rules, etc., with Differential Diagnosis and Treatment. By ORVILLE HORWITZ, B.S., M.D., Demonstrator of Anatomy, Jefferson Medical College; Chief, Out-Patient Surgical Department, Jefferson Medical College Hospital. 3d edition. Very much Enlarged and Rearranged. 91 Illustrations and 77 Formulæ. 12mo. *No. 9 ? Quiz-Compend ? Series.*
Cloth, 1.00; Interleaved for the addition of Notes, 1.25.

Pye's Surgical Handicraft. A Manual of Surgical Manipulations, Minor Surgery, Bandaging, Dressing, etc., etc. With special chapters on Aural Surgery, Extraction of Teeth, Anæsthetics, etc. 208 Illustrations. 8vo. Cloth, 5.00

Swain's Surgical Emergencies. New Edition. Illus. Clo., 1.50

Walsham. Manual of Practical Surgery. For Students and Physicians. By WM. J. WALSHAM, M.D., F.R.C.S., Asst. Surg. to, and Dem. of Practical Surg. in, St. Bartholomew's Hospital, Surgeon to Metropolitan Free Hospital, London. With 236 Engravings. *See Page 2.* Cloth, 3.00; Leather, 3.50

☞ *See pages 15 and 16 for list of ? Quiz-Compends ?*

THROAT.

Mackenzie on the Throat and Nose. New Edition. By Morell Mackenzie, M.D., Senior Physician to the Hospital for Diseases of the Chest and Throat; Lecturer on Diseases of the Throat at the London Hospital, etc. Revised and Edited by D. Bryson Delavan, M.D., Prof. of Laryngology and Rhinology in the N. Y. Polyclinic; Chief of Clinic, Department of Diseases of the Throat, College of Physicians and Surgeons, N. Y.; Sec'y of the Amer. Laryngological Assoc., etc. Complete in one volume, over 200 Illustrations, and many formulæ. *Preparing.*

—— Diseases of the Œsophagus, Nose and Naso-Pharynx, with Formulæ and 93 Illustrations. Cloth, 3.00; Leather, 4.00

"It is both practical and learned; abundantly and well illustrated; its descriptions of disease are graphic and the diagnosis the best we have anywhere seen."—*Philadelphia Medical Times.*

Cohen. The Throat and Voice. Illustrated. Cloth, .50
James. Sore Throat. Its Nature, Varieties and Treatment. 12mo. Illustrated. Paper cover, .75; Cloth, 1.25

URINE, URINARY ORGANS, ETC.

Acton. The Reproductive Organs. In Childhood, Youth, Adult Life and Old Age. Sixth Edition. Cloth, 2.00
Beale. Urinary and Renal Diseases and Calculous Disorders. Hints on Diagnosis and Treatment. 12mo. Cloth, 1.75
Holland. The Urine, and Common Poisons. Chemical and Microscopical, for Laboratory Use. Illustrated. 2d Edition. Cloth, .75
Ralfe. Kidney Diseases and Urinary Derangements. 42 Illustrations. 12mo. 572 pages. Cloth, 2.75
Legg. On the Urine. A Practical Guide. 6th Ed. Cloth, .75
Marshall and Smith. On the Urine. The Chemical Analysis of the Urine. By John Marshall, M.D., Chemical Laboratory, Univ. of Penna; and Prof. E. F. Smith, PH.D. Col. Plates. Cloth, 1.00
Thompson. Diseases of the Urinary Organs. Eighth London Edition. Illustrated. Cloth, 3.50
Tyson. On the Urine. A Practical Guide to the Examination of Urine. With Colored Plates and Wood Engravings. 6th Ed. Enlarged. 12mo. Cloth, 1.50
—— Bright's Disease and Diabetes. Illus. Cloth, 3.50
Van Nüys, Urine Analysis. Illus. Cloth, 2.00

VENEREAL DISEASES.

Hill and Cooper. Student's Manual of Venereal Diseases, with Formulæ. Fourth Edition. 12mo. Cloth, 1.00
Durkee. On Gonorrhœa and Syphilis. Illus. Cloth, 3.50

☞ *See pages 15 and 16 for list of ? Quiz-Compends ?*

NEW AND REVISED EDITIONS.

?QUIZ-COMPENDS?

The Best Compends for Students' Use in the Quiz Class, and when Preparing for Examinations.

Compiled in accordance with the latest teachings of prominent lecturers and the most popular Text-books.

They form a most complete, practical and exhaustive set of manuals, containing information nowhere else collected in such a condensed, practical shape. Thoroughly up to the times in every respect, containing many new prescriptions and formulæ, and over two hundred and thirty illustrations, many of which have been drawn and engraved specially for this series. The authors have had large experience as quiz-masters and attachés of colleges, with exceptional opportunities for noting the most recent advances and methods. The arrangement of the subjects, illustrations, types, etc., are all of the most approved form, and the size of the books is such that they may be easily carried in the pocket. They are constantly being revised, so as to include the latest and best teachings, and can be used by students of any college of medicine, dentistry or pharmacy.

Cloth, each $1.00. Interleaved for Notes, $1.25.

No. 1. **HUMAN ANATOMY,** "Based upon Gray." Fourth Edition, including Visceral Anatomy, formerly published separately. Over 100 Illustrations. By SAMUEL O. L. POTTER, M.A., M.D., late A. A. Surgeon U. S. Army. Professor of Practice, Cooper Medical College, San Francisco.

Nos. 2 and 3. **PRACTICE OF MEDICINE.** Third Edition. By DANIEL E. HUGHES, M.D., Demonstrator of Clinical Medicine in Jefferson Medical College, Philadelphia. In two parts.

PART I.—Continued, Eruptive and Periodical Fevers, Diseases of the Stomach, Intestines, Peritoneum, Biliary Passages, Liver, Kidneys, etc. (including Tests for Urine), General Diseases, etc.

PART II.—Diseases of the Respiratory System (including Physical Diagnosis), Circulatory System and Nervous System; Diseases of the Blood, etc.

** These little books can be regarded as a full set of notes upon the Practice of Medicine, containing the Synonyms, Definitions, Causes, Symptoms, Prognosis, Diagnosis, Treatment, etc., of each disease, and including a number of prescriptions hitherto unpublished.

BLAKISTON'S ? QUIZ-COMPENDS ?
CONTINUED.

Bound in Cloth, $1.00 **Interleaved, for Notes, $1.25**

No. 4. PHYSIOLOGY, including Embryology. Fourth Edition. By ALBERT P. BRUBAKER, M.D., Prof. of Physiology, Penn'a College of Dental Surgery; Demonstrator of Physiology in Jefferson Medical College, Philadelphia. Revised, Enlarged and Illustrated.

No. 5. OBSTETRICS. Illustrated. Fourth Edition. By HENRY G. LANDIS, M.D., Prof. of Obstetrics and Diseases of Women, in Starling Medical College, Columbus, O. Revised Edition. New Illustrations.

No. 6. MATERIA MEDICA, THERAPEUTICS AND PRESCRIPTION WRITING. Fifth Revised Edition. With especial Reference to the Physiological Action of Drugs, and a complete article on Prescription Writing. Based on the Last Revision of the U. S. Pharmacopœia, and including many unofficinal remedies. By SAMUEL O. L. POTTER, M.A., M.D., late A. A. Surg. U. S. Army; Prof. of Practice, Cooper Medical College, San Francisco. Improved and Enlarged, with Index.

No. 7. GYNÆCOLOGY. A Compend of Diseases of Women. By HENRY MORRIS, M.D., Demonstrator of Obstetrics, Jefferson Medical College, Philadelphia. *In Press.*

No. 8. DISEASES OF THE EYE AND REFRACTION, including Treatment and Surgery. By L. WEBSTER FOX, M.D., Chief Clinical Assistant Ophthalmological Dept., Jefferson Medical College, etc., and GEO. M. GOULD, M.D. 71 Illustrations, 39 Formulæ. Second Enlarged and Improved Edition. Index.

No. 9. SURGERY. Illustrated. Third Edition. Including Fractures, Wounds, Dislocations, Sprains, Amputations and other operations; Inflammation, Suppuration, Ulcers, Syphilis, Tumors, Shock, etc. Diseases of the Spine, Ear, Bladder, Testicles, Anus, and other Surgical Diseases. By ORVILLE HORWITZ, A.M., M.D., Demonstrator of Anatomy, Jefferson Medical College. Revised and Enlarged. 77 Formulæ and 91 Illustrations.

No. 10. CHEMISTRY. Inorganic and Organic. For Medical and Dental Students. Including Urinary Analysis and Medical Chemistry. By HENRY LEFFMANN, M.D., Prof. of Chemistry in Penn'a College of Dental Surgery, Phila. A new Edition, Revised and Rewritten, with Index.

No. 11. PHARMACY. Based upon "Remington's Text-book of Pharmacy." By F. E. STEWART, M.D., PH.G., Quiz-Master at Philadelphia College of Pharmacy. Second Edition, Revised.

Bound in Cloth, $1. Interleaved, for the Addition of Notes, $1.25.

☞ *These books are constantly revised to keep up with the latest teachings and discoveries, so that they contain all the new methods and principles. No series of books are so complete in detail, concise in language, or so well printed and bound. Each one forms a complete set of notes upon the subject under consideration.*

www.ingramcontent.com/pod-product-compliance
Lightning Source LLC
Chambersburg PA
CBHW031348160426
43196CB00007B/775